A CHANCE TO SERVE

A memoir of service in the
Marine Corps and Marine Corps Reserve

BRIAN POWER

A Chance to Serve: A memoir of service in the Marine Corps and Marine Corps Reserve

Produced and printed
by Stillwater River Publications.

Visit our website at
www.StillwaterPress.com
for more information.
First Stillwater River Publications Edition
ISBN: 978-1-955123-01-3
Library of Congress Control Number: 2021905575
1 2 3 4 5 6 7 8 9 10
Written by Brian Power
Published by Stillwater River Publications,
Pawtucket, RI, USA.
Publisher's Cataloging-In-Publication Data
(Prepared by The Donohue Group, Inc.)

Names: Power, Brian, 1946- author.
Title: A chance to serve : a memoir of service in the Marine Corps and Marine Corps Reserves / Brian Power.
Description: First Stillwater River Publications edition. | Pawtucket, RI, USA : Stillwater River Publications, [2021]
Identifiers: ISBN 9781955123013
Subjects: LCSH: Power, Brian. | United States. Marine Corps--Officers--Biography. | United States. Marine Corps Reserve--Officers--Biography. | Marines--United States--Biography. | LCGFT: Autobiographies.
Classification: LCC VE25.P691 A3 2021 | DDC 359.96092--dc23

*The views and opinions expressed in this book
are solely those of the author
and do not necessarily reflect the
views and opinions of the publisher.*

Cover Art:

The cover portrait was done by (then) Corporal Andy Vonada whom I had the pleasure of leading when I was a Reserve Captain platoon commander with 'H' Company, 2nd Battalion, 25th Marine Regiment.

This memoir is dedicated to all who serve
and have served honorably in our armed forces.

Table of Contents

A CHANCE TO SERVE

THE MARINES

A Need for Self-Discipline

1.

PARRIS ISLAND

I was trained at Parris Island,
In the land that God forgot,
Where the sand is fourteen inches deep,
And the Sun is piping hot.

When I die and go to heaven,
To St. Peter I will tell,
Sir, another Marine reporting in,
I've served my time in hell.

Unofficial last verse of the Marines' Hymn

"Hey guys! Look! Palm trees. I bet this place is really nice."

Naïve recruit looking through the window as the Greyhound bus pulled onto Marine Corps Recruit Depot, Parris Island, SC, near midnight, end of August 1968

Let me give you a little background on Marine Corps Recruit Depot, Parris Island, South Carolina. MCRD Parris Island is a flea ridden sand barren that the Marines had been using for recruit training since 1915. If you are not a Marine recruit, given appropriate bug repellant and sunscreen, it might be a tolerable, if not pleasant place to spend a little time. If you are a Marine recruit, Parris Island is not exactly hell, but you can feel its heat, smell the sulfur, and feel Satan's terror from there.

As the buses pulled up to receiving barracks and halted, our driver headed straight for what appeared to be the statue of a large Marine. The Marine didn't move as the Greyhound bus pulled up nose to nose with him and came to a stop just before the windshield bent the brim of the Marine's Smokey the Bear hat. Our driver got off the bus like he was shot from a cannon. The Marine came to life, immediately got on the bus, and standing at the junction of the front seats and the exit, in a voice conditioned by too many unfiltered cigarettes and years of straight whiskey, he announced his presence with a bellowing authority:

"IF YOU ARE CHEWING GUM YOU WILL SWALLOW IT IF YOU ARE SMOKING YOU WILL PUT IT OUT YOU WILL GETOFF MY BUS IN TEN SECONDS FLAT AND FALL IN ON THE YELLOW FOOTPRINTS YOU WILL NOT TOUCH ME YOU WILL NOT TOUCH ME YOU WILL NOT TOUCH ME..."

(On the street near where the bus had parked, there were 80 sets of yellow footprints lined up in four rows of twenty. Each pair of footprints had the feet with heels together and toes pointed outward at a 45-degree angle.)

And boy did we empty that bus fast. Guys were climbing over seats and through windows to avoid this menacing apparition. Once on the footprints we were met by two other Drill Instructors who surrounded us like a lynch mob, screaming at the top of their lungs, **"GET ON MY FOOTPRINTS GET ON MY FOOTPRINTS GET ON MY FOOTPRINTS...ARE YOU EYEBALLING ME STUPID? BECAUSE IF YOU ARE I'M GOING TO TEAR YOU A NEW ASSHOLE RIGHT IN FRONT OF THESE OTHER GIRLS!!!"**

Despite the lore that I had read in Marine combat histories, and despite all the stories that my Community College former Marine friends told me -- and that I had ignored at my peril -- I suddenly realized that life, as I had once known it, ended in that moment.

Once assembled on the footprints we were screamed at relentlessly for a few more minutes before being herded into the Recruit Receiving Barracks where there were several rows of very long and narrow tables.

In a bellowing monotone we were told, **"YOU WILL EMPTY YOUR POCKETS AND OVERNIGHT BAGS COMPLETELY. YOU WILL PLACE ALL ITEMS ON THE TABLE IN FRONT OF YOU. A DRILL INSTRUCTOR WILL SEARCH YOUR BELONGINGS FOR**

CONTRABAND. CONTRABAND CONSISTS OF CIGARETTES, CHEWING GUM, DRUGS, KNIVES, BRASS KNUCKLES, AND GUNS. THOSE WE WILL CONFISCATE. YOU WILL SEND ALL JEWELRY EXCEPT FOR WEDDING RINGS BACK HOME."

I thought that was a most interesting inventory to be concerned with. I was expecting the bit about chewing gum, jewelry, and cigarettes, but knives, brass knuckles, and guns??? Yet, sure enough, a couple of switchblades were confiscated. I imagine that long experience had taught the Marine Corps that not everyone they recruited was a choirboy.

We were not the only group brought in that evening. Three other groups of recruits arrived about the same time we did and were being given the same processing treatment in other areas of the building.

After our warm welcome, we were separated into groups of 75 or so and distributed among four separate barracks. My group was brought topside (upstairs -- we had to learn a whole new lexicon of Naval service terms such as topside/upstairs, bulkhead/wall, scuttlebutt/drinking fountain, deck/floor, ladder/stairs, hatch/door, etc.) where we were told we would live for the next four nights. We had arrived on the Friday night of Labor Day weekend and wouldn't be assigned our permanent Drill Instructors until Tuesday morning. AAA definitely did not plan that travel itinerary.

During the night there were Marines who patrolled our barracks to ensure that no one got wandering fever.

Each morning at 6:00 AM, and again at noon, and later at 5:00 PM, three anonymous Drill Instructors (different ones every time) would gather us up and herd us to the mess hall where the food was surprisingly good (I later found out that the best chow in the Marine Corps was served at the two boot camps at Parris Island and San Diego as well as at the infantry training commands where we were sent right after boot camp). We could take anything we wanted on our trays, but we had to **"EAT EVERY GODDAMN CRUMB SWEETHEART, OR ELSE!"**

Somehow we got through the long weekend, mostly by passing around some trashy paperback books, mostly Mickey Spillane type novels, that a few of our more literate brethren had wisely stashed in their bags.

I once read that Marine Boot Camp was carrot and stick on the scale of the Passion Play at Oberammergau. It was an apt description, as I was about to learn.

On Tuesday morning following Labor Day, right after morning chow, bright and early, we were told to line up at the tables again. Our permanent Drill Instructors had arrived. We were told that when our names were called we were to **"RUN, NOT WALK, TO THE HATCH (doorway) WHERE A DRILL INSTRUCTOR WILL CHECK YOUR NAME OFF AND WRITE A NUMBER ON YOUR HAND. THEN YOU WILL WALK, NOT RUN, DOWN THE LADDER (steps), WHERE YOU WILL RUN, NOT WALK, TO THE YELLOW FOOTPRINTS. YOU WILL STAND THERE AT ATTENTION UNTIL THE ENTIRE PLATOON IS ASSEMBLED. YOU WILL NOT TALK. YOU WILL LOOK NEITHER LEFT NOR RIGHT, AND GOD HELP YOUR ASSES IF YOU LOOK DIRECTLY AT A DRILL INSTRUCTOR!!!"**

I believed this last admonishment referred to the "Burning Bush" syndrome of the Old Testament, the details of which, I figured, would be covered later.

That first morning was the beginning of an amazingly well orchestrated chaos. One DI called our names, the next one at the hatchway verified the name and wrote a number in magic marker on the backs of our hands (my number was 57), and we ran out the door to the chaotic screaming of the drill instructors, to wit, **"YOU WILL STAND EXACTLY ON MY FOOTPRINTS!!! YOU WILL NOT LOOK ANYWHERE BUT STRAIGHT AHEAD!!! YOU WILL NOT SMILE!!! YOU WILL NOT TALK!!"** And so on and so forth.

Once all 75 of us were gathered outside, we were herded to the next destination, which was the barbershop staffed by five masters of their craft. We stood in line and filed in five at a time. There, the team of barbers – artisans, members of a guild, no doubt -- who, too modest to sign their works, in a matter of 20 seconds, sheared our once prized late 1960s hair down to the nub. We all looked pretty stupid afterwards standing in our now four-day-ripe civilian clothes with shaved heads. Heck, it was 1968, and no one, but no one anywhere on planet earth, except at Parris Island and San Diego Recruit Training Battalions, had a haircut like that.

Next stop was the supply shed where we were each issued a seabag (a green canvas bag about 3.5 feet tall and maybe 2.0 feet in diameter) and a full complement of uniforms and web gear (packs, belts and suspenders from which to hang the accoutrements of war), underwear and socks. The clothing was fitted precisely by a Lance Corporal who eyeballed each of us and made a SWAG (Scientifically Wild-Assed Guess) as to our sizes. If any of us were stupid enough to say that an item was too large or too small, said Lance Corporal told us we would either grow into it or that we would lose weight. I am sure these guys all landed successful retail sales jobs at Mens' Warehouse and Joseph Bank when their Active Duty service ended.

Most everything we were issued went into the seabag, save one cover (a Marine green peaked hat, similar enough to describe as 'like a baseball hat' to the untutored reader); three sets of underwear and socks, one of which we immediately put on; two sets of utilities (green dungaree-type of working uniform that was to be our main uniform for the next ten weeks), one of which we put on; and two pairs of boots, one of which we put on.

Everything we weren't wearing (except our civilian clothes) went into the seabag.

Before we left this site, we packaged up our civilian clothing and addressed it home. It would be a long time before we wore civvies again. God help our parents who had to open those reeking sets of civilian clothes.

Next stop, the armory, where we were each issued an M-14 rifle (a 9.5 pound item that would become like an appendage over the next 10 weeks). Along with the weapon we were issued ammunition magazines and pouches to hold them, a first aid kit, and rifle cleaning gear. Everything but the rifle was put in the seabag. We were (shall I say strongly?) admonished that if we ever dropped our rifle, it would take a team of medical specialists weeks to remove it from a certain portion of our anatomies. My rifle number was immediately imbedded into my brain-housing group (to use the rifle-oriented parlance of the day), and which I was compelled to recite, along with my serial number (2477262) at the top of my lungs many times in the first weeks of training.

Our newly forming platoon also stopped at the PX (Post Exchange) to get "health and comfort items." These consisted of shaving gear, baby powder, a galvanized bucket, a scrub brush, shoe polish, cloth diapers (I'll explain that one later), clothes pins, a large bar of brown scrubbing soap, liquid starch, toothpaste, and possibly a couple of other items. These items were charged to our pay accounts, where the money was deducted. Everything we accumulated was stuffed into either our seabags or the galvanized bucket.

There may have been other stops on our itinerary, but you get the idea. We were burdened down with this weight and staggering along in "herd formation" (in God-awful coastal South Carolina heat and humidity) with our DIs screaming, **"YOU SILLY PEOPLE ARE TOO SLOW!!!"** Finally, they herded us up the four or five steps into our new home, an "H" shaped, white-shingle sided wooden barracks that had been built in the early days of World War II.

(This building, if you had an arial view of it, and several others like it, each housed four recruit platoons on the first and second floors of the 'H'. The middle bar of the 'H' contained the drill instructors' offices as well as other office spaces along the middle bar for the platoon series commanders.)

We were ordered to find our number (remember the number written on our hands?). The numbers were written on the deck (floor) in chalk in front of each rack (bunk). I found number 57 and waited there for the next order. Even numbers got the bottom rack, odd numbers got top. We were to put our rifles on our racks. We were to turn around and face the passageway (aisle) and raise our 60-pound sea-bags over our heads at full arm extension. We were to keep our sea-bags over our heads until the Drill Instructors got tired.

Let me tell you about Parris Island weather the first week of September 1968. It was a steam bath. I mean IT WAS A FUCKING STEAM BATH! It was incredibly hot outside with both temperature and humidity in the high nineties, and in the barracks, with the heat and sweat of 75 idiots polluting the atmosphere, it was Dante's Inferno, only *"New and improved"* as they say on Madison Avenue.

So there we stood in silence while our DIs yelled and screamed and ran up and down the barracks accusing us of being put on the planet just to piss them off. Finally one guy down at the other end of the barracks passed out and crashed to the floor with his sea-bag. I figured that they would ease up a bit now, but the DIs surrounded him and screamed, **"IS HE DEAD YET? IS HE DEAD YET?"**

This was something my sophisticated northeast educated suburban brain was not processing very well. Very quickly a medical Corpsman showed up and revived the poor soul and took him away. We never saw him again.

The next hour or so lives in my memory like no other. The Drill Instructors informed us that we were now members of Platoon 2017, and if and when they called for "School Circle," we were to freeze at attention, shout as loud as we could, "SCHOOL CIRCLE, AYE, AYE, SIR!" The next several minutes were spent by our shouting 'SCHOOL CIRCLE, AYE, AYE, SIR' as loud as we could until the DIs were satisfied that we understood the instruction.

We were then to wait for the DIs to give the execution command and say, "DO IT!" At that point, we were to move our stupid asses as fast as we could move them to the center of the squad-bay and stand at attention until we heard the command, "READYYYYYYYY, SEATS!" On the command of 'READY,' we were to cross our arms over our chests and on the command of 'SEATS,' drop to a cross-legged "Indian" position on the deck (floor).

The Drill Instructors, Staff Sergeant J. J. McDowell (the senior DI), and Sergeants W. D. Rivers and D. A. Crall (the assistant DIs), gave us the appropriate commands, and, because of their disappointment with our response, we needed a few dozen more rehearsals to **"GET IT FUCKING RIGHT, MAGGOTS!!!!."** Once we had mastered this basic command, they gave us three classes that burned away any illusions that I had about anything.

The first class was on how to escape from Parris Island. This was for those of us who just couldn't handle the pressure of trying to become a Marine. The first escape option was the quarter-mile long main causeway leading from the mainland onto the base. That corridor was not only patrolled regularly, but there was a guardhouse at the end of the causeway with a sentry who had orders to shoot to kill any recruit trying to escape.

Next, there was the quicksand that would suck us down to the center of the earth.

Finally, there was the swimming option, which involved sharks, barracuda, and riptides.

So, there was no escaping Parris Island.

The next class, if you have a bizarre way of thinking, was almost a logical sequel. The class was on how to commit suicide.

The Drill Instructors would not tolerate any of this nicking the wrists bullshit. Uh, uh. No Sir! That just meant that you were looking for attention. And nobody in THEIR Marine Corps drew attention to themselves. Marines were part of a team – not individuals. Nooooooo! If you were going to do it, there was only one option available, and that was to dig one's razor blade into the wrist and slice deep all the way down to the crook of the elbow. That way, we were assured, they knew we were serious, even though there would be not a damned thing they could do about it.

Jesus H. What had I gotten myself into?

The final class presented by the Drill Instructors was actually the one that should be taught to everyone everywhere — in Fortune 500 corporation Board Rooms, federal and state governments, churches, schools, athletic organizations — I mean everywhere. This class was simple and to the point. I don't have the exact quote memorized, but it went something like this: "WE WILL NOT TOLERATE ANYONE USING THE FOLLOWING WORDS: SPIC, KIKE, MICK, WOP, DAGO, GUINEA, POLOCK, KRAUT, NIGGER, CHINK OR ANYTHING ELSE THAT YOU MORONS MIGHT USE TO DESCRIBE ANYONE WHO IS DIFFERENT FROM YOU, BECAUSE ***YOU WILL NEVER QUESTION THE COLOR OF A MARINE'S SKIN, OR THE WAY HE WORSHIPS GOD, WHO PULLS YOU OUT OF HARM'S WAY***!!!!!"

After all the horror that had just been presented, I suddenly felt reassured.

And these three Drill Instructors meant it, as I will relate a little later on.

After we had a moment to absorb this, they explained to us that the Series Lieutenant (we were one training platoon out of four that made up a "Series") would be giving the next class and that when they announced him, we were to snap to attention faster than immediately, and then resume our lotus position when the lieutenant said, "As you were." We were informed that officers were like gods and were to be treated as such. We also rehearsed this movement a couple of dozen times until we got it more or less right.

The Series Lieutenant, First Lieutenant Wells, appeared, and we dutifully, as best we could with the minimum of practice, snapped to standing attention. The Lieutenant gave us the "As you were" command, we dutifully dropped back into the lotus position, and his presentation began. He explained in colorful detail the Uniform Code of Military Justice and the delightful array of crimes and misdemeaners on display therein for which the mal-performing Marine could be sent to Portsmouth.

(Portsmouth Naval Prison in New Hampshire, at that time, was the Naval Service's equivalent of Devil's Island. It was a "Red Line" brig where outside of your cell a red line was painted every six feet. You had to ask permission of the shotgun-armed Marine guards on a catwalk above for permission to cross the red line. A few years later, in a Pentagon cost-cutting measure,

Portsmouth was closed and the festivities were moved to Fort Leavenworth, Kansas.)

Boy this place was going to be fun.

After Lieutenant Wells departed, the DIs asked those of us who had been to college to stand up. About a third of the platoon had some college, anywhere from one semester to four years. They asked which of us had been English majors. A few of us raised our hands. They ordered the remainder to be seated. They looked closely at each of us who remained standing and then selected me. **"YOU WILL BE THE PLATOON SCRIBE. YOU WILL TAKE CARE OF ALL TRAINING RECORDS. WE WILL TEACH YOU WHAT YOU NEED TO KNOW. YOU WILL NOT FUCK UP."**

I wasn't sure just what kind of honor had just been bestowed on me, but I was to find out very soon.

Next we were taught how to make up our racks (bunk beds), and how to do it most efficiently using our bunkmates to help, because Marines functioned as a team, and not as individuals. First there was the fart sack (I'm not kidding), a long heavy-weave cotton bag that fit over the mattress and protected it "FROM YOUR SLIMY BODIES." Then the bottom sheet that had to be fitted in just a certain way with "hospital corners." That was when one of my fellow recruits, one Herbert Malcolm Waldo Brigham III, late of one of the Ivy League schools, said, "Sir, our maid always said…" He of course got no further. **"YOUR MAIDS WERE ALL WRONG!!! YOUR MOTHERS WERE ALL WRONG!!! THERE IS ONLY ONE FUCKING WAY TO FOLD MARINE CORPS SHEETS AND ONLY ONE WAY TO MAKE UP A MARINE CORPS RACK AND THAT IS THE MARINE CORPS WAY!!!"**

We were starting to get the point. Never look directly at a Drill Instructor or you would be turned into a block of wood. No escape from Parris Island. Don't even think about suicide. Don't screw up or it's gonna be military prison. Racism was forbidden — if you came in with certain attitudes, they were going to be knocked out of you. Our mothers were wrong (not to mention our maids).

* * *

A LIFE OUT OF CONTROL

As I finished my second year at Norwalk Community College in 1968, I was becoming more and more frustrated with my life. I was working two part-time jobs and spending everything I earned on partying and not having any fun. I was drinking too much. As for the future, I couldn't see further than the opening of a beer can, and I most definitely had no goals for my college education. I didn't know why I was in college. For heaven's sake I was now completing my third year of college work and because of my sloppy study habits had only accumulated just a little under two-years worth of college credits.

Yet during my two years at Norwalk Community College, I encountered several young men, mostly in their mid- to late-twenties, who had served in the armed forces and were now taking advantage of the GI Bill to go back to college and get a degree. Those who were former Marines had filled me with their lore as well as the sagas of their recruit training experiences.

In short order, I was hooked. Actually that was easy. I had been reading about the military since I was about twelve, and I knew that one day when I did join a service, it would be the Marines. These guys at Norwalk Community College just pushed me over the edge.

Finally I knew it was time. One day in late spring of 1968 I simply went down to the Marine recruiter in the basement of the South Norwalk, Connecticut Post Office. The Gunnery Sergeant behind the desk looked formidable. I said, "I know the Marines have four-year and three-year enlistments that guarantee specialized schools as an inducement, but what does the two-year option guarantee?"

He answered, "A hard time, a rifle, and a chance to serve your country."

"I'll take it," I said.

It was 1968, and the dead and casualty numbers from Vietnam were frightening. I was 22 years old and said to myself, "Take charge of your life because you are heading for the toilet." This was not the most rational decision people of my age were making. I needed no one's permission. I signed up on June 6, 1968, and, at the recruiter's excellent suggestion, I took the delay option so I didn't have to report until August 29th, thereby avoiding the worst that summer could offer at Parris Island, SC. I was instructed to carry a minimum amount of gear: a toothbrush, toothpaste, shaving gear, a couple of changes of underwear, and a couple of changes of socks. When I got to Parris Island, I would be sending all that back (except the toothbrush and toothpaste) along with the clothes I wore down there.

I was the first of my family, maternal or paternal, to join the military since my great-grandfather Power had, according to my cousin, enlisted in the British army to fight in the Crimean War. As family legend has it, he came back injured and angry at the callous way the Brits used the Irish and other colonials as front-line troops. Despite his injuries he refused a British pension for his efforts. In recent generations both my maternal and paternal grandfathers were too old to serve in WWI, and my father was too old to serve in WWII.

My father couldn't believe what I had done, but finally agreed that it was my decision. My mother said almost nothing, but I later realized that she was terrified by the prospect of me coming back from southeast Asia in a body bag like so many others who had served. My brother and sister said little, but I knew they were just as concerned.

I got on the train in South Norwalk on Friday, August 28, 1968. The first stop was about twenty miles up the coast of Connecticut where I would get my final physical exam at the New Haven Armed Forces Induction Center. After that unpleasant experience that can only be categorized with the term 'Group Indignity,' I joined up with all the other recruits from the region. That afternoon, having connected with the other New England region recruits, we got back on the train and headed south to Parris Island, SC, home of the Marine Corps recruit Training Depot..

On the train I met Curtis Brown who only three years before had been one of the starting five on Norwalk High School's state championship basketball team. He had married Jackie Robinson's daughter (Jackie Robinson was the legendary Hall of Fame baseball player for the Brooklyn Dodgers, who not only was the first black Major League player in the modern era, but who also was a great champion of Civil Rights). But the marriage turned out to be a disaster and they were soon divorced. I shared the entire trip to South Carolina with him and we hit it off.

When the train pulled into Charleston, SC, in the evening the next day, we were met by some Marines who moved us quickly to the Greyhound buses outside. A short time later, actually sometime near midnight, we were pulling onto the well-lit causeway leading onto the Marine Corps Recruit Depot, Parris Island, when a voice from the back of the bus said, "Hey, look. Palm trees. I bet this place is nice."

* * *

Cleanliness may have been the principle ethic at Parris Island. We were taught to get on our hands and knees and scrub every square inch of the wooden deck with hand-held stiff wire brushes until the wood gleamed (I have never been in a cleaner environment than boot camp barracks). I also read somewhere that for the Marines, cleanliness was the only ethic. Most of my fellow recruits had never cleaned up after themselves in their lives. Mama, and in a few cases, the maid, always did it. Now we were all getting an education in this essential virtue. We gained a new respect for our mothers and maids. And we were taught Marine Corps History, and how to march, and the manual of arms (Left Shoulder Arms, Right Shoulder Arms, Port Arms, Order Arms, etc, etc., etc.).

We each had a Marine Corps Guidebook, a loose-leafed book that fit into the breast pocket of our utility uniform. From that guidebook we had to memorize a jillion facts. First were the Eleven General Orders. ('One: I will take charge of this post and all government property in view. Two: I will call the corporal of the guard in any case of fire or disorder, etc.') I was fortunate to have a good recruiter who gave me a copy of the general orders and made me memorize each one before I left for Boot Camp. Drill Instructors would take delight in challenging recruits to recite, say, the fourth general order. When the recruit failed to do it properly, it would be push-ups until the DI got tired. I made a good early impression by being able to spout them at will.

Let me give you an appreciation of the typical Boot Camp day.

At 0500 (5:00AM – we use the 24-hour clock) the Drill Instructors would interrupt our slumber by throwing metal trash cans around the barracks and beating on them with swagger sticks (military batons carried by NCOs and officers at that time but soon to be phased out) as they flicked on the lights and shouted, **"RISE AND SHINE, MAGGOTS. IT'S ANOTHER GRAND AND GLORIOUS DAY IN THE MARINE CORPS! EVERY DAY A HOLIDAY, EVERY MEAL A BANQUET, AND EVERY DUTY STATION A COUNTRY CLUB!"**

At the first syllable of their reveille, we had better be standing at attention at the end of our racks with our sheets in our hands, or God help our sorry asses. The sheets in our hands ensured that no wiseass would sleep on the floor to avoid making up his rack in the morning. Our first task was to count off –

ONE! TWO! THREE! etc., up to the last man, Private Williams, who was number 75. He would then take one step forward and shout, "SIR, SEVENTY-FIVE MARINE CORPS RECRUITS ON DECK, SIR!"

The DIs had assigned one additional duty to Private Jones, a slow talking black kid from the hills of Georgia. He was to be the Platoon 2017 Weather Man for the duration. Every day after count-off, the duty DI would say, "Hey Mr. Weather Man. What's the weather going to be?" And Jones would take one step forward and say the same thing every day for the next seventy days: "My name is Captain Jones, and the weather I have found, for Beaufort, Parris Island, and the area all around. It's gonna be hot."

That essential ritual completed, we had five minutes to go to the head (bathroom), pee, wash our hands, brush our teeth, shave, and be in PT (Physical Training) uniform – red PT shorts, green T-shirts, green wool socks and combat boots – and standing at attention at the end of our racks. Try to imagine a life-sized picture of 75 Marine recruits trying to use seven urinals, seven toilets (all exposed, by the way, with no walls for privacy -- a situation to make even the mildly fastidious blanch), and seven sinks, all at once. Upon the completion of morning ablutions, we had another five minutes to make up our racks, do an initial shape up of the barracks, and fall into formation outside the squad-bay. Then it was pull-ups, sit-ups, squat thrusts, and a three-mile run in twenty-three minutes to warm up. As soon as we got back, and this was now sometime around 6 AM, we had three minutes to shower, get into our utility uniform and fall out for march to chow.

At the chow hall, we lined up by twos and marched in, removing our covers (hats) as we entered the mess hall (Marines are never covered indoors unless 'under arms' – i.e., wearing a canvas duty belt). I remember one totally absurd moment (Boot Camp was filled with them) when one of my platoon mates accidentally stepped inside the chow hall without removing his cover. He was immediately surrounded by a Drill Instructor who made the recruit apologize to each of the food groups. "The Private is sorry, potatoes, the Private is sorry vegetables, etc." You had to be there.

Once inside the mess hall we grabbed our trays, side-stepped through the chow line **(YOU'D BETTER BE SIDE STEPPING THROUGH MY CHOW LINE, MAGGOT!!!),** loaded up with food, and in the first few weeks of Boot Camp, never got to sit and eat it. With a Drill Instructor screaming in our ears, we had to shovel the food into our mouths as we headed from the chow line to the rear hatch (doorway) and then scrape our trays, stack them and get into formation to go back and finish cleaning the barracks. It was during these leisurely moments waiting for the platoon to form up that we were to study our guidebooks. Inside of these guidebooks resided the history, customs, and traditions of the Marine Corps. In every spare moment when we weren't otherwise purposed, we flipped the books open and read and memorized tons of facts about the Corps. The Marine Corps had learned over it's existence that the lore and legend of its exploits were critical to the bonding of new recruits to the collective consciousness of the Corps.

(A brief note here about Boot Camp operations in the 1960s. On our first day there were about 80 of us. The DIs went through the ranks and culled out the severely overweight recruits. They were assigned to "Fat Body Platoon" where they would be put on a special conditioning and diet program and then re-cycled into other platoons later when their weight and physical conditioning merited their safe participation in the regular Boot regimen. Today that can't happen. In the modern – sensible – Marine Corps, the recruiter is held responsible for sending a well conditioned recruit to Boot Camp. The Marine Corps doesn't need to create another layer of training units to condition overweight recruits.)

Our DIs singled out five fat bodies for designation to conditioning platoons, but they decided to keep a few others who were obviously overweight, but who, because of the look in their eyes, indicated that they could hack the regular program.

One of those saved for continuation in our platoon was my bunkmate, Albert Prunier. Albert was from Massachusetts and stood about six foot two or three. He weighed, at the beginning of Boot Camp, 257 pounds; he was big and soft. He and a few others designated as fat bodies would always have to line up at the end of the chow line **(FAT BODIES TO THE REAR OF THE CHOW LINE!),** and their food selection and intake were closely monitored by the DIs.

For the first several weeks, Albert was only allowed a minimalist meal at each feeding. Breakfast was one bowl of a Cheerio equivalent (purchased at the lowest cost by the Marine Corps procurement section). Said cereal was to be covered with only skimmed milk – sugar was forbidden. Orange juice was mandatory. Lunch consisted of the equivalent of salad with oil and vinegar dressing, one slice of bread, and a small glass of skimmed milk. Dinner was more or less the same. By the end of Boot Camp, Albert was pretty much lean and jacked up and weighed in at 197 pounds.

(This was an interesting phenomenon. The overweight guys dropped pounds like crazy, and guys like me, who had entered Boot Camp at five foot ten and a-half and a skinny 151 pounds, came out at 172 pounds, also jacked up.)

Once the platoon was fed and formed up in three squads, we were marched back to the barracks.

At the barracks we were commanded to give the place a thorough cleaning, like we didn't do this every day since we arrived. I was generally exempt from cleaning the barracks because of my duties as Platoon Scribe. In that role, I was to prepare in triplicate on 3x5 cards, the daily training schedule, which each Drill Instructor tucked into the inside sweatband of his Smokey the Bear hat.

The schedule varied slightly from day to day because on some mornings we were scheduled for early chow, and on others we ate late (i.e., at 0630). If, because of the rotating chow schedule we hadn't done our PT, we would then change into PT gear after chow (red shorts, green T shirts and boots) and fall out in formation for a three-mile run. One way or another – PT before chow or PT after chow -- It was now only about 0630. The rest of the civilized world had not even gotten out of bed yet.

The PT regimen was physically demanding, to say the least. The first order of business was to complete the three-mile run. That was followed by push-ups and squat-thrusts (squat-thrusts, otherwise known as "bends and motherfuckers") was an exercise performed by dropping to a squat with your hands on the ground between your knees and then, resting on your hands, thrusting your feet backward to a push-up position, doing a push-up, snapping back to a squat, then to the standing position; the ultimate result of doing **MANY AND MANY OF THEM, MAGGOTS!!!** was that one's thighs ended up feeling like they were on fire from the inside). We also did pull-ups and bent-knee sit-ups with a partner anchoring our feet, and then after being marched back to the barracks, we were given 5 minutes to shower (remember, seventy-five guys, seven shower heads).

We immediately changed into our utility uniform and fell out in formation again. The rest of the day was spent in both classroom instruction on subjects such as Marine Corps History; the Workings of the M-14 Rifle; the care, maintenance and firing techniques associated with the .45 Caliber Pistol; and Combat First Aid. Additional familiarization classes included Land Navigation and a range of other subjects.

A note on our study of Marine Corps history: The Marines are obsessed with imbedding the story of its origins, its major battles and wars, and the individual officers and enlisted men who exemplified our standard. This lore was essential to our ultimate emergence into the collective consciousness of the most storied military organization on the planet. The inculcation of Marine Corps lore was essential to the production of the final product: One fully trained and motivated U.S. Marine.

When we weren't in classrooms, we were outside on the parade deck (grinder) learning how to become a team by learning platoon close-order drill.

All this was spiced up with the occasional visit to the medical clinic where we lined up and ran a gauntlet of medical Corpsmen who had these pneumatic injection guns that inoculated us from every disease then known to man. That was a new experience for all of us. We would line up and walk through a room that might have seven or eight Navy Medical Corpsmen lined up on each side of us, all armed with pneumatic injection guns. It wasn't uncommon for us to have sore upper arms in those first weeks of Boot Camp. After receiving an ungodly number of injections in both biceps, we would leave the clinic, and then we of course still had to carry our weapons and do pushups.

Other times it might be dental clinic that intruded on our warrior training. As you might well imagine, there were some members of the platoon (well, more than half) who had never had the pleasure of meeting a dentist in person. You gotta hand it to the military, you could have arrived from Harvard with three years of college, or you might have been a subsistence farmer from Alabama with a seventh-grade education, but you all got first rate, quality health and dental care.

Throughout the training, every minute of every hour of every day was planned to the nth detail. And through it all, if we were in an upright position, we were carrying our rifles at "Port Arms," which means our right hand held the stock at waist level and our left hand the barrel group at left pectoral level, with the rifle diagonally across and six inches from our chests.

(Tom Seaver, the Hall of Fame pitcher for the New York Mets, was a Marine and went through Boot Camp at San Diego, CA. He said carrying that rifle was the one thing that built up his arm strength to become a Hall of Fame power pitcher.)

With appropriate breaks for chow at noon and dinner (20 minutes each), we had a full day. Typically after dinner we would have more drill practice. At 2000 (twenty-hundred, or 8:00PM) we had our free time for an hour. This was a time for mail call as well as a lax period where we could clean our rifles; spit-shine our boots (this was what the diapers were for; we tightly wrapped the diaper around out index finger, wetted it with water poured into the lid of the shoe polish tin, rubbed our finger into the shoe polish, and rubbed it onto our boots in small circular motions until, over the weeks, the boots built up a mirror layer of gleam). Using all this free time we would write letters home, read our mail, polish our brass; take a shower; and do anything else that didn't involve leaving the barracks, eating snacks, or having fun.

At the beginning of our one-hour break, we had mail call. That was another interesting ritual. At precisely 2000, we recruits all stood at attention in front of our racks. The Duty Drill Instructor for the evening, stationed at the center of the squadbay, would call out the recruit's name and the recruit would take one step forward, execute a right-face or left-face, and then double-time up to the DI and say, "Sir! Private So-and-So reporting for mail call as ordered, Sir!" And the DI would hold out an envelope. The recruit would then slap the envelope between his two palms, the Drill Instructor would let go, and the recruit would take one step backward, do an about face, then double-time back to his rack and stand at attention. No one was allowed to open his mail until all the mail had been distributed. We lived for the letter from home, or from a sweetheart, or from a friend, or a wife.

Then the Fire Watch was posted. This was where members of our platoon were assigned in two-hour segments, to patrol up and down the barracks during 'lights out' to ensure that no one had decided to "go over the hill," and that no inappropriate behavior was going on. Once the Fire Watch was posted, the DI would order us to prepare for sleep. We would stand at attention in front of our racks, and he would give the order, **"GET IN YOUR RACKS,"** which we had to do with great alacrity or we would have to rehearse it a few dozen times. Then he would say, **"WHEN THESE LIGHTS GO OUT, I'D BETTER HEAR YOUR EYEBALLS SLAMMING SHUT."** And then there would be instant, merciful sleep.

Thus passed a typical day for Platoon 2017.

The Marine Corps also looked out for our spiritual welfare. On Sundays there were religious services and we were "encouraged" to attend. It was made perfectly clear that if we didn't go to Catholic Mass or Protestant services, the Drill Instructors would find something for us to do in the barracks. Such an encouragement produced loads of religious converts. So every Sunday the Catholics would form up at one time and march to Mass, and when we returned, the Protestants would march off to whatever it was that they did on Sundays.

Don't get the idea that the rest of us just hung out while one group was in Chapel. Sundays were for other major activities. The first order of business was to do laundry. Now before you form a picture of a bunch of Boots standing around some machines waiting for the spin cycle to finish, let me tell you what laundry day was really about.

Outside each of the World War II barracks there were long metal tables with drains and a line of elevated faucets every few feet. We would take our sweat-soaked utilities, socks and underwear, our galvanized buckets, our bars of brown soap, and our scrub brushes to these tables and scrub out our clothes by hand. As we finished each piece, we'd wring it out and drop it in our galvanized buckets full of clear water for the rinse cycle. When the entire "load" was done, we put it through the "dry" cycle, which consisted of wringing everything out by hand, and hanging everything up on clotheslines that the Corps benevolently provided for our use.

A couple of hours later, after scrubbing every inch of the barracks on our hands and knees (for the seventh time that week), we would get our laundry from the clotheslines and we'd each have about 10 minutes with the sole platoon iron to make the uniforms look reasonably presentable.

The rest of Sunday was pretty low key. Even the DIs needed a break. Time to catch up on letter writing, boot polishing, brass shining, rifle cleaning, and the rest.

Thus passeth the Sabbath.

Throughout Boot Camp there were a series of events upon which each platoon was graded. These were the Savings Bond Drive, Initial Physical Fitness Test, Preliminary Essential Subjects (the stuff in the manual that we memorized), Initial Close Order Drill (about three weeks into the program), Rifle Range Scores (sixth week), Final PFT (week 10), Final Essential Subjects (ninth week), Final Close Order Drill Competition (tenth week), Final Inspection (final week).

Each platoon in the series competed against the other three platoons of our 'Series' in these events. For each event "won," the platoon got a colored streamer for its guide-on (the small flagstaff that held our platoon flag and which was carried at the front of platoon formation by the Platoon Guide).

Every platoon was expected to have 100% participation in the Savings Bond program. We all signed up with the admonition "You don't have to do this, but…" This was a pretty good program, actually. You could sign up for as big a bond as you could manage, but with what we were paid (in 1968 Privates were paid about $105 per month), we all signed up for the $25 Bond program. For this, $18.75 per month was withheld from our pay, and we were, in fact, enrolled in an enforced savings program.

Every platoon in the series got 100% participation in this event.

So Platoon 2017 got its first streamer for 100% Savings Bond participation. (I continued the bond program throughout my four years of active duty and saved $5000 that I was able to use as a down payment on our first house in Harrisburg, PA, which was financed by the GI Bill.)

Then there was the initial subjects test. By the end of the third week or so, we were expected to have a mastery over certain basic Marine Corps subjects, including some of the Marine Corps history. I believe we needed something like a 75% or 80% platoon average score to get the streamer.

Platoon 2017 got a streamer for this event for having the highest composite scores in the Series.

Initial Close Order Drill was a major event. The preliminary competition around the third or fourth week between the four platoons in our series was an event of major note as emphasized by our Drill Instructors. There would be a final drill competition toward the end of our training in the tenth week. In the initial drill competition we were, of course, a little intimidated. I mean, we had only been there three weeks, and here we were trying to do some of the basics of close order drill in front of a group of judges culled from the ranks of drill instructors from other battalions.

We won it and got another streamer.

We went out to the rifle range for our fifth and sixth weeks to learn the necessary disciplines of shooting a rifle properly. This was the Marine Corps' bread and butter. Every Marine is a rifleman, regardless of whether he (or she) is an admin clerk, supply pogue (pogue is a derisive term applied to anyone who had a rear-echelon job), aviator, or grunt (infantry). It was also an interesting psychological break. There was a lessening of the harshness of training so we could concentrate on mastering our weapons. We still went out in every spare moment and learned more and more about close order drill and the myriad of variants within that discipline, and we also studied our guidebooks and memorized thousands of facts about the Marine Corps, but most of every day was spent on the rifle range.

The first week we learned the fundamentals of shooting from a steel-gray eyed Marine sniper. In outdoor classrooms he taught us how to hold the rifle in the proper position with our cheekbone firmly resting against our thumb that was wrapped around the wooden stock of the M-14 rifle. That position was called the "spot weld." In particular, we learned how to get a proper sight picture (fuzzy bulls eye, sharply defined front sight blade), and how to get into the four basic shooting positions (standing, kneeling, sitting and prone). This was what was referred to as "snapping-in." We 'dry-fired' our weapons over and over with the firm but quiet encouragement of the instructors. I later learned that the Drill Instructors deliberately eased off on us to remove our tension so that we would be more relaxed when it came to shooting for qualification.

I was also to learn that all of our successes (or failures) were part of the Drill Instructors' evaluations. I was also later to learn that in most of corporate America, if an employee or employees failed, it was their fault. Their managers were rewarded for getting new replacements, not for improving employee performance. The Marine Corps was different. If a Recruit failed at a task, it was his Drill Instructor's responsibility.

This was my introduction to Leadership.

The second week at the rifle range we got our "dope." This is an interesting term that has nothing at all to do with what was going on elsewhere in the '60s. The term had everything to do with how it was used before the Hippie generation. Before the '60s, the word 'dope' meant information. "So what's the latest dope?"

We discovered that each rifle, though manufactured by the same company and supposedly identical, was actually different from every other rifle by marginally small measures. Additionally, each Marine was different from all other Marines. Our size and shape determined how the rifle was positioned in each of our hands. To get the proper "dope" on a rifle, you first fired it on the 1000-inch range to find out what was "zero" for your rifle. By this I mean that a shooter, in order to "zero" his rifle, needed to find out on the 1000-inch range how many clicks of elevation and windage were required to be turned on the sights to make a bullet strike a bullseye.

Ergo, a given rifle, because of its manufacture, and because of the shape of the Marine holding it, might require two clicks of elevation and three of "right" windage to strike a bullseye. That was "zero" dope for that particular rifle.

Next we got to do live fire practice at targets 200, 300, and 500 yards away. We practiced in the standing, sitting, kneeling, and prone positions. Here we learned to factor in additional windage and elevation on our sights, as the trajectory of a bullet drops an inch for every hundred yards of distance from the target. We learned to determine windage by looking at the pennants at either end of the firing line to estimate how strongly the wind was blowing. The wind would also alter the strike trajectory of a bullet a certain distance, depending on how far from the target it was fired. We adjusted our sights from the "zero" position either right or left an appropriate number of clicks to get the proper windage for our rifles as it applied to each of the shooting positions at each of the ranges. We recorded all of this data in our shooter log-books.

At 200 yards we might not have to adjust the elevation of our rifles very much, if at all, from the values we determined on the 1000-inch range, but as we moved to the 300 and 500-yard lines, our sight elevation would have to be adjusted a certain number of clicks, along with windage adjustments.

The final data was the "dope" on our individual rifles. Now all that remained was to fire for qualification.

On Thursday of the second week at the range, we fired for practice, and then made our final adjustments to the rifle sights and logged everything into our books. On Friday we fired for "qual."

So here I was, this basic eastern liberal former English major whose toughest challenge to date was dealing with the folks in a suburban *cul de sac* world, and I was attempting to qualify with a rifle for the first time in my life. As luck, skill, and solid Marine training would have it, at 500 yards – that's 1500 feet – I hit 'Black' seven out of ten times.

I had the second highest score in our platoon with a 223 overall score. I was a Marine Expert Rifleman, and boy did this northeastern liberal feel great about it.

Eventually on my dress uniforms I would wear the crossed rifles of an expert marksman. I am proud to say that with an exception one year, throughout my career, even as a lieutenant colonel, I still shot expert with both rifle and .45 cal. pistol.

And yet, to this day, I never developed an interest in either owning weapons or firing them for any reason whatever. And I am totally content with that. I loved the preparation for competition and was simply thrilled by this unexpected achievement.

Platoon 2017 ended up with the highest percentage of qualifying scores for our series, and we won the marksmanship streamer for our guide-on.

We spent another week at the rifle range with about half the platoon off doing mess duty and the other half doing other work details. Again, our spare time was filled with more drill and more learning. We were starting to get pretty good at drill, and the DI's started teaching us "trick drill," that is, fancy movements and turns that are not part of the Marine Corps Manual for Close Order Drill. The platoon pride as a result of our achievements in the past few weeks was beginning to show.

We also spent a tremendous amount of time on a detailed cleaning of our weapons after a week of solid shooting.

We drilled intensively this 'slow' week after rifle qualification, with pil-low cases over our heads, on disassembling and re-assembling our rifles. The DIs kept a stop watch on us while they would slowly walk from one end of the barracks to the other as we went through this blind drill. Before long every man in the platoon could strip and re-assemble a rifle blind in about 15 seconds as we might have to do in combat, at night and with no ambient light. I had a feeling that this would look good on my resume.

An interesting incident happened that week that reflects back on our "race relations" instructions of the first day. I had mentioned that the first day we met our new Drill Instructors, they "selected out" some of the "fat bodies" to go to a conditioning platoon before they could be accepted into the training regimen. Other fat bodies at an earlier date had been selected for similar con-ditioning by other Drill Instructors. By the end of our second week of training, we had picked up a few of these former fat bodies who were now in good enough physical shape and within the weight requirements to continue with Boot Camp training.

One of these former fat bodies was Private A, an amiable enough blond fellow who did not have trouble fitting in with our platoon culture. Un-fortunately, because of his former overweight status, he had not been privy to the race relations lesson the rest of us had received on the first day. So while we were in that final week of work details at the rifle range barracks, one day when I happened to be in the Drill Instructors' hut doing paperwork, there was a loud banging on the DI's hatch. BANG! BANG! BANG! "SIR, PRIVATE W REQUESTS PERMISSION TO SPEAK WITH THE DRILL INSTRUCTOR SIR!"

Sgt. Rivers was the duty DI that afternoon, and he said, "Get in here, Maggot!" Private W properly squared himself in front of the Drill Instructor's desk and reported. Sgt. Rivers growled, "Speak!"

"Sir, the Drill Instructors always said to report when someone used an improper racial term, Sir."

Now I was very interested. Sgt. Rivers, who was both white and from Georgia, and a real hard-ass, got real quiet, and said in a low voice, "What happened, Pvt. W?"

W, who was a nice black kid from farm country somewhere in the south, said, "Sir, Private A called the Private (we had to refer to ourselves in the third person) 'Nigger.'"

Sgt. Rivers was giving off some dangerous signals, like a throbbing of a vein in his temple and a rising of redness in his face, but he said quietly, "W, just go back to your rack and continue cleaning your rifle, and I will handle this."

W, seeing the look in Sgt. Rivers' eyes, was picking up the vibe, too. He said, "Aye! Aye, Sir." Took one step backward, did an about face, and got the hell out of there.

"PRIVATE A!!!! REPORT TO THE DRILL INSTRUCTOR IMMEDIATLY!!!" bellowed Sgt. Rivers.

A showed up like someone was chasing him, did the appropriate banging on the hatch, and reported to the Drill Instructor. As he was standing there trembling in front of Sgt. Rivers' desk, Rivers looked at me and said, "Get lost, but not too far."

I immediately got out of there and stood at attention a little way down the passageway. The door to the drill instructors' hut closed with a definitive click. From where I waited, I could hear the sickening sounds of Private A ricocheting off one bulkhead after another, while at the same time Sgt. Rivers was, shall we say, providing A with a period of instruction on race relations.

After a few minutes, things got quiet. Then I hear Rivers say, **"YOU WILL APOLOGIZE TO PRIVATE W AND YOU WILL DO IT**

RIGHT NOW AND IF I EVER HEAR OF YOU SAYING SUCH A THING AGAIN I WILL PERSONALLY STRANGLE YOU!!!"

I considered that to be one of the privileged conversations I would not share with anyone. Not the Commanding General, not the Inspector General of the Marine Corps, not the Commandant himself. As the statute of limitations has probably long passed, I feel that I can share the moment with you.

A came almost crawling out of the Drill Instructor's hut and ran past me down the passageway right to W's rack and I heard him saying, "I'm sorry man, I mean I am really sorry. It was stupid of me. It will never happen again. I mean it, really." And W, stunned by A's disheveled appearance, said, "It's okay, man, its cool."

I was just processing this when Sgt. Rivers blasted out of his hut and growled at me, "Get your ass in here, Maggot, and clean up this mess you made!!!"

I went into the hut, and it looked like a hurricane had hit. And I immediately got to work cleaning up the mess I had made.

I'd like to provide a further note, for all those who have never had the pleasure of experiencing Marine Boot Camp. Legend has it that at Marine Boot Camp there is rampant brutality. In my time, the late 60s, I knew it did go on, as my later conversations with Marines from other recruit training platoons of my era affirmed. But it was not a sanctioned institutionalized practice. The Marines have what is called the SOP or Standard Operating Procedures. The SOP forbids physical contact with a Recruit other than to correct a position (of Attention, Parade Rest, Port Arms, and the like). I assure you, it never happened in my platoon. Marine Drill Instructors have a unique bond with their platoons. If the bond is built properly, the threat of violence is enough. And if that bond is built properly, when there is an occasional, and justified, incident of violence, the platoon will act as one and swear to a man that it never happened. In my entire stay at MCRD Parris Island, I never saw one incident of brutality by Drill Instructors in my platoon. The only incident that I could consider was the incident with Private A.

I never saw it.

And it was richly deserved.

Two reasons for this lack of physical punishment in my platoon were, 1) that we had so many high school graduates – two-thirds of the platoon, people who had accomplished something and not dropped out; and 2) we had three hard-assed but decent leaders who set an example for us all. Our having a platoon with such a high percentage of high school graduates and recruits with anywhere from one semester to four years of college was luck of the draw, really. We had a smarter than average recruit platoon for the time (1968), and our Drill Instructors knew it and didn't need to go around thumping everyone just to get their attention.

The average GCT (General Classification Test) score for our platoon was 105. The GCT pretty much paralleled the IQ test scores. Our platoon's average GCT Test score of 105 was slightly above average compared with other platoons at that time. Our highest GCT score was something like 180 and our lowest was 90 (128 for yours truly because I knew you wanted to know). Factored into this is the fact that our Hispanics had limited English speaking skills and our African Americans came, in many cases, from Jim Crow under-funded and poorly performing schools in the rural south. But they all were a smart group. We were a smart platoon.

* * *

(One more note on Marine training: Up until we went to the All-Volunteer Force [AVF] in the mid-70s, the Marine Corps tended to get more than its share of "jail or the Corps" recruits. So many enlistees were high school – or even junior high school -- drop-outs who resented authority and had never accomplished anything. They got to Boot Camp and challenged the DIs' authority as they challenged authority all their lives. When many of them got to Vietnam, they caved under the pressure and performed below par.

(Another factor in 1968 was that even though the Standard Operating Procedures [SOP] forbade physical punishment, it did happen. Fortunately it rarely got totally out of control. With the new AVF in the mid-70s, the Marine recruitment target was a 95% high school graduation rate. Recruiting now focused on bringing recruits aboard who had played team sports or had other accomplishments in high school. The other 5% were required to score even higher on the screening tests than their high school grad counterparts. That being said, the Corps also set up new psychological screening standards for Drill Instructors. These screens were designed to weed out those with latent authoritarian characteristics who couldn't handle the pressure of 90-hour work weeks. Bottom line: better recruits and better DIs, and a real drop off in incidents of brutality.)

* * *

It was now time for Platoon 2017 to return from our temporary rifle range residence to our original barracks at mainside. We marched back feeling a little "salty," a Marine term for cockiness. Our Drill Instructors disabused us of the feeling quickly enough. We still had a lot of learning to do and a few more competitions to face before we could remotely be considered human beings, not to mention qualified to graduate as actual Marines. The tempo now stepped up a notch.

I need to throw in this story before I forget it. Albert Prunier was my bunkmate. He was one of the originally designated 'marginal' body types that our DIs had decided to retain because he showed some intensive desire. As I mentioned earlier, when he checked in, at six foot-two he weighed 257 pounds.

Albert was as good hearted and fine a person as anyone would ever hope to meet. We got along famously from the beginning, always looking out for each other. One day, however, there was an incident when we were doing pugil stick training that is worthy of note. Pugil sticks are padded sticks that look like big Q-Tips on steroids and are about the same weight as an M-14 rifle (nine and-a-half pounds). Pugil stick fighting is a simulated way of preparing for the really possible wartime experience of fighting an enemy at close quarters after you've run out of bullets. Each contestant in our pugil stick drills wears a football helmet and a steel protective cup strapped on the outside of the uniform to protect the genitals.

We had done pugil stick training on several occasions, and each time we were paired against a recruit of similar size. I won some fights and got knocked on my ass a few other times. On this one occasion the Drill Instructors were feeling whimsical and decided that rather than pair off recruits of equal size, we should have bunkmates fight each other.

I was not looking forward to that, because Albert had me by almost 4 inches and, at the time, about 50 pounds.

On one side of the ring formed by the platoon, SSgt. McDowell was threatening Albert with mortal pain if he did not take me out. On the other side of the arena, Sgt. Rivers was strongly advising me that if I lost this match, I might never see tomorrow morning, and if I did, I'd be watching sunrise from a hospital room and eating meals through a straw.

Albert and I approached each other and I noticed he was winking at me. I realized that he wanted us to put on a show for the Drill Instructors and maybe see if we could get a draw. At the same time, Sgt. Rivers voice was going through my head. I winked back. So we sparred for a short time, then when I saw him drop his guard, I caught him on the side of the helmet with a Horizontal Butt stroke and knocked him ass over tea kettle.

I received much praise from Sgt. Rivers and accolades from the platoon for defeating a much larger opponent, but I still felt bad when I saw Albert cranking out endless push-ups for SSgt. McDowell. To Albert's eternal credit, he said it was no big deal, and that if he had been thinking with a clear head, he would have clocked me right from the start.

As platoon scribe, I was privileged to be exposed to a lot of private Drill Instructor conversations. These guys changed uniforms three to five times a day back in the Drill Instructors' hut in order to appear totally "squared away" at all times. It's a fundamental Marine leadership principle to always set the example. So after each training evolution such as close order drill or PT, I would have a few minutes to get some more scribe work done while they changed uniforms and the rest of the platoon secured rifles or modified their gear for the next training evolution.

One particular day, right after noon chow call, the platoon was getting ready to go out to the "grinder" (the parade ground) once again for drill practice, but this time to learn a particular drill movement that was not that easy to do. It required several rifle movements while on the march. The platoon was to march in precise unison while going from port arms to left shoulder arms, back to port arms, and then to right shoulder arms then back to port arms again. We had mastered each of these movements one at a time from the stationary position, now we were going to get them all at once on the march.

The Drill Instructors were strategizing on how they would handle our basic failure to do this. Historically the majority of recruit platoons (with the usual high percentage of educational drop-outs) had an enormous amount of trouble getting this particular part of close order drill correct. The Drill Instructors' standard "school solution" to the problem was to criticize the platoon, call them babies, and then return the platoon to the barracks for a one hour nap. The purpose of this thinking was to humiliate and motivate. In the past, this method had worked at getting platoons mad enough to really focus on getting it right when they got the second chance.

In the real world it is a bankrupt approach to leadership.

What our DIs hadn't yet fully appreciated was the fact that we were a unique platoon. Based on our education levels, as a unit we were both smart and teachable.

They realized that I was there in the office as they were talking and they cautioned me to say nothing about their plans. That was the last thing on my mind. So we fell out in formation with our rifles and marched over to the grinder. The DIs gave us the proper instruction, we marched forward, and, lo and behold, Platoon 2017 got the commands right on the first try. And the second try. And on every try afterward. So they kept us out for an hour or so, marched us back to the barracks, and ordered us to spend the next hour cleaning our rifles and polishing our boots.

Back in the Drill Instructors' hut, I resumed my administrative duties while the DIs changed uniforms once again. They were floored that we got it right away and now we had a one-hour hole in the training schedule. They were nearly delirious that we seemed to take to close order drill so well. I remember one of them asking me how the platoon did it, and I responded, "Sirs, this is a smart platoon. At least one-third has had some college, that group and another third are high school graduates, and the other third are smart even though they don't have a high school education."

It was not the first, nor the last time that I was allowed into their confidence. I always respected their private conversations, and I never betrayed a confidence. They became very comfortable talking freely in my presence, and I was honored by that fact.

There was another interesting incident that happened about five weeks into the training when we were still out at the rifle range that confirmed for me what master psychologists and observers of human nature our Drill Instructors were. Our platoon had a recruit named Private B. For the first five weeks of Boot Camp, B was like wallpaper. You knew he was there because you kept hearing his name called at roll call. But he made no impression. I don't think he ever spoke. He just absorbed himself into the collective consciousness of the platoon.

As usual, one night at 2000 we had mail call. SSgt. McDowell happened to be the DI on duty that night. He called out B's name, and then nothing happened. We were all standing at attention, but started to look down to where B's rack was. SSgt. McDowell again called out his name, and slowly B walked up to the Drill Instructor and said, "Private B reporting for mail call as ordered, Sir". But he said it in such a quiet voice, it was hard to hear him. And of course he was not doing the drill in the strong-voiced military manner we had been taught.

SSgt. McDowell then said in a gentle voice, "This is your first letter, isn't it?" And B answered in the affirmative. McDowell said, "I want you to go into my hut (the DI office) and read your letter. I'll be in shortly." He finished mail call and went into his office. As the Platoon Scribe, this was also my "free time" to do my additional duties of laying out the schedule for the next day, so I went into the office, too. I saw SSgt. Mc Dowell sit on the Drill Instructor's duty rack and put his arm around B's shoulders and start to talk gently to him. It turned out that B was an orphan who had been reluctantly raised by an abusive relative. When he was in high school he had no friends, save for one girl who took an interest in him. It was from her that he received this first letter. Mc Dowell said to B that he should see the DI first thing in the morning, and that he wanted B to talk to a doctor he knew at the clinic, just so that B would feel better. B said, "All right, Sir. I'll do it."

SSgt. McDowell asked me to say nothing about this incident to anyone, and I agreed wholeheartedly. A couple of months later, when I was doing post-Boot Camp infantry training at Camp LeJeune, I was near the PX one day when I heard, "Hey, Power!" I turned around and there was B. He said he had seen a psychologist at the clinic at SSgt. McDowell's recommendation, and had been through some additional counseling at Camp LeJeune, and he was pulling his life together. I told him how happy I was to hear that. We had a few more words, and he moved on. Sadly, I never saw him again.

But what stood out in my mind was the incredible insight these DIs had. I was to be continuously astonished.

One other incident that happened around the same time is locked into my consciousness forever. During the first week of Boot Camp there were ten thousand administrative things going on in order to get us absorbed into the system. One of these was a battery of tests called the CLEP (College Level Equivalency Proficiency) Test. This was an intelligence test that determined your college level equivalency. If you scored above the 35th percentile on all five tests, had one year of college, and you were at least 19 years of age, you met the first level of qualification for being eligible for the Enlisted Commissioning Program (ECP). This was where you could start the process that would allow you to apply for Officers Candidate School.

Back at Mount St. Mary's College I had signed up for the Marine Corps officer training program that would be conducted between my Freshman and Sophomore years and again between my Junior and Senior years. By flunking out at the end of my freshman year, I had blown my chance for that program. But Boot Camp offered another shot at becoming an Officer of Marines.

At mail call one night, again SSgt. McDowell, our senior Drill Instructor, was on duty. I had already received a letter from home and a letter from Kathy, and suddenly my name was called again: "Private Power!" I reported to the Drill Instructor and slapped my palms together on the envelope, but the Senior DI didn't let go. He said, "Open it." I looked at the envelope and realized that this was official correspondence from the Marine Corps. I'm sure that SSgt. McDowell's initial reaction was defensive. Did this recruit send a letter to the Inspector General? Was he to be investigated?

I opened the letter and read the first few words, then handed it to SSgt. McDowell.

I had held some hope that I might be considered for the ECP, but this letter was formal notification that because one of my grades on the CLEP test was below the 35th percentile (it was 34 point something!!) that I could not be considered for the program.

SSgt. McDowell read the letter, looked at me for a second, and said, "They're a bunch of God-damned fools."

Never had I been let down so well, and never had I been so determined to do better.

Week by week, in spite of our occasional individual screw-ups Platoon 2017 kept getting better at everything it did.

A brief note here about the sand fleas. From a recruit's perspective at Parris Island, sand fleas were a protected species. The EPA would disagree. In reality, sand fleas are nothing more than flying teeth. We were forbidden to slap at them, brush them away, or do anything that would disturb them. As Sgt. Crall once said, "They have their own Marine Corps and they need to be fed."

So for the entire ten weeks of Boot Camp, we were forbidden to do anything to them. And they were everywhere. We developed a lot of self-discipline by not slapping them at every opportunity. In our final week, we were allowed to slap at them at will.

We still had a few weeks remaining, and one of our big tests coming up in the last week was the final test on all we had memorized in the Marine

Guidebook, everything we had learned in classes about Marine Corps history and traditions, everything about the M-14 rifle, and probably a lot more.

What we needed in order to prepare for the big test was a major tutoring effort. Our DIs had "encouraged" us to come up with a tutoring plan, but they did not want to know the details. They needed deniability. Our squad leaders, all well-chosen by our DIs, paired each of us who had some college with a high school grad, and as a team we tutored the drop-outs and the limited-English speaking in our spare time. My special student was S R, a farm boy from way out in the hills of North Carolina. I once asked him where he was from and he said, "Plum." I said, "Where the heck is that?" and he said, "Plum back in the woods." S was a great guy, but academics were not his strong suit.

The problem was that we had no spare time, so what we did was wait until after "lights-out" and held classes under blankets that we hung from the upper bunks and draped around the lower bunks. We conducted the classes with flashlights illuminating the guidebooks and drilled our students so that after a couple of weeks they could spout the correct answers to any of the questions.

The final test was actually part of an overnight exercise called "Eliot's Beach." (They don't do Eliot's Beach anymore. They replaced it with a grueling 54-hour no-sleep exercise called "The Crucible." Glad I went through when I did.) Eliot's Beach started with a long forced march of many miles, with full packs and rifles, around and through this 'flea ridden sand barren' we had called home for the past nine weeks. At various stops along the way, usually at monuments and plaques, we got more lectures on Marine Corps history and traditions. (I can't even begin to tell you how important that is to Marines.)

Finally in the late afternoon we arrived at Eliot's Beach. We were given instruction on how to create a comfortable, temporary field home. Soon we had constructed our shelter halves, had chow, and before settling in for the night, sat around a bonfire and listened raptly as our Drill Instructors told us a bunch of "sea stories" of Marine Corps battles both historical and recent. One story has a wonderful irony attached to it. Our Series Gunnery Sergeant (Platoon 2017 was one platoon in a series of four platoons), was a man named M. This Company Gunnery Sergeant was one total hard ass with a voice that could bring down thunder. As our DIs related, in Vietnam one night, his unit was attacked while they were sitting, seemingly in a non-combat area, around a camp fire. A Viet Cong soldier charged into the clearing and our Gunny said, calmly, "Somebody please kill this bastard.""

The irony of the story happened about a year after I graduated from Boot Camp. This same Gunny, this one man force, was court martialed and sent to prison for homosexual behavior. Thank you for your service. Fortunately we have become a little more enlightened in recent years.

The next morning after chow we were broken down into small teams and sent to different stations around the bivouac area. At each station each of us had to answer questions about everything we had been taught at Parris Island. The questioners were, for the most part, Drill Instructors from other battalions. They recorded our names and graded our answers. S and a couple of other of our after-hours students were with the team I was on.

And each of these guys got at least passing grades on the test, and some of them actually excelled. By the end of the day, when each team had rotated through all of the stations, it turned out that ours was the only platoon in the series to have 100% of the recruits pass the test.

Platoon 2017 picked up another streamer for the guide-on.

We now had two major hurdles to overcome: the final Drill Competition and final Battalion Commander's inspection.

Drill Comp is the big one because it is the one test in which it is really the Drill Instructors who are being graded. Our preparations for this one were spectacular. We each gave up one set of clean utilities to one of our junior DIs, Sgt. Rivers, who washed and tumble dried them at his home. (And this is another extraordinary example of what these Dis did at their own expense after spending 90 hours a week with their recruits.)

For our individual preparation, each recruit soaked his cover in water, wrung out, and stretched it over a cylindrical plastic frame called a cover block. Then each of us took clothes-pins, anchored the cover on its block, and pulled down the cover to smooth out the wrinkles. At that point we painted our covers with a coating of liquid starch. They were left to dry and were not to be moved until the day of drill competition.

Over the course of our training, each evening we worked on our day-to-day gear, and that included spit-shining our extra pair of boots until we could use them as mirrors. We cleaned our rifles until they gleamed.

We scrubbed our extra belts, which we had never worn, with Crest toothpaste (no kidding) to bleach them white, and we polished our brass belt buckles to the point of radiance.

Our days were filled with endless practice on all the drill moves that could be required of us in the competition — flanking movements, column movements, oblique movements, and any number of other movements. For the competition, our senior DI, SSgt. McDowell, would be drilling us, but he wouldn't know what movements we would be required to execute until we arrived at the parade deck.

Finally the day of drill competition came, and we were wired. We took turns helping each other get dressed so that none of us would bend over and put creases and wrinkles in the uniforms. We each stood on a footlocker while our bunkmate held out our trousers to step into. We took turns tying our partner's bootlaces while he stood on the footlocker.

At the appropriate time, we fell out in formation and began the march of a few blocks to the parade ground. As we marched, SSgt. McDowell began singing the Battle Hymn of the Republic in time with our cadence, as a way of relaxing us. When he finished the first verse, for some reason I can't explain, I began singing the third (I think it's the third) verse:

In the beauty of the lilies Christ was born across the sea,
With a glory in His bosom that transfigures you and me;
As He died to make men holy,
Let us live to make men free,
His truth is marching on.

It was a very emotional moment as we sailed onto the parade deck like a ship confident of its course. Platoon 2017 stood off to the side and watched as two of the other platoons went through their paces. We were third to go, and after watching the others go through their motions, we knew we were better. Finally it was our turn.

SSgt. McDowell was given the list of movements we were to execute, and he moved us onto the gridiron. In the middle of a couple of movements he was instructed to bring us to a halt while the evaluators walked around us to see if we were still properly aligned with each other. Then we would continue. Our 75 pairs of boot heels were hitting the pavement at exactly the same time. Our rifle movements were crisp and sharp, with each of our hands slapping the rifle stocks at exactly the same moment. We were a musical band in total sync.

Platoon 2017 put another streamer on its guide-on. We had won the battalion drill competition and our Drill Instructors were completely ecstatic.

The trophy they had won was a large wooden pedestal with a bronzed pair of boots on it. Sgt. Rivers and Sgt. Crall carried the trophy back to the barracks while marching alongside SSgt. McDowell and the platoon. When we got back to the company street in front of our barracks, and were standing there at attention, none of the Drill Instructors knew quite what to say. Finally Sgt. Crall said, "Make a head call!" (This was a rare privilege which we usually only got right before chow, or during our free time from 2000 to 2100 at night.)

We responded "Aye, Aye, SIR! Make a head call, Sir." Then he said, "Do it!" and we broke ranks and made for the head (bathroom) whether we needed to go or not. When we came out and got in formation again, he said to make another head call. We were all totally crazy with the release of tension. It might not mean much to you who have not gone through it, but this was a spectacular moment for Platoon 2017.

Now there was only one major event remaining — Final Inspection by the Battalion Commander himself. I think that event was two days before graduation. Again, it was one requiring meticulous preparation. This inspection required us to be in our Dress Green uniforms (or Alphas as the Marines call them), which we had never worn, and which we had been fitted for by a sharp-eyed Lance Corporal early in our training.

We went for our final fittings about a week or ten days before the inspection, and of course most of us had greatly changed shape from the first weeks of the program. I had gone from a relatively skinny 151 pounds to a pretty muscular 172 pounds. My bunk mate had gone from a blubbery 257 to a jacked up 197 pounds. There was much tailoring to be done. At last, the day before the inspection, our uniforms were delivered. We had already spent weeks spit-shining our dress shoes so that they had a mile-deep mirror shine. Our rifles were cleaned one final time, then wrapped in cotton sheets to protect them.

The only thing that remained was to go over every square inch of the uniform and remove the Irish Pennants. An Irish Pennant was any stray thread that peeked out from the corner of a pocket, or the hem of the blouse, or the collar, or the lapels. The term 'Irish Pennant' originated in the British navy of the 1800s. There, the original connotation was that an "Irish pennant" was a threadbare or frayed rope or halyard, or an unsecured, untidy rope end hanging without purpose from a ship's rigging.

The British never thought much of the Irish.

Again we dressed each other carefully to avoid creases and wrinkles. We retrieved our rifles and fell out for formation.

After inspecting all the rifles and uniforms and asking questions of individuals, the Battalion Commander and his inspection team declared that Platoon 2017, with the fewest discrepancies of any platoon in the series, was to be the Honor Platoon of the Series.

Platoon 2017 put the final streamer on the guide-on, and now we carried the title of "Honor Platoon." (As an additional reward, following our graduation, each of our DIs was promoted meritoriously.)

Graduation now was two days away, and we would finally be called Marines.

Our parents were all encouraged to come down for the ceremony, and Mom, Dad, and Elaine made the trip. After they arrived on the day before graduation, they were briefed by the Public Affairs team and told that they would be able to meet their son for a few hours that afternoon in the Visitors Center. I, along with several other of my platoon mates, headed for the Visitors Center to await our parents' arrivals. When I saw them coming up the steps, I rushed out to greet them and they almost walked by me. As I said, I had changed quite a bit. I now had this really short haircut, if you could call it that. The sides and back of my head were "high and tight" which means shaved nearly to the skin. And the hair on top of my head was only about a quarter inch long. Plus my face and arms were deeply tanned, my torso had filled out with muscles, and I was a lot bigger.

We had a great reunion and I introduced them to many of my friends. The following day they would watch from the bleachers along the parade deck as I became a Marine.

The day before I graduated, after visiting my parents and sister, I learned that I was going to be one of the seven from our platoon to be promoted meritoriously to Private First Class. SSgt. McDowell called me down to the Drill Instructors' hut and in the presence of sergeants Crall and Rivers read my citation to me. Then the three of them performed the time-honored Marine tradition of taking turns "pinning" my chevron on by lining up on either side of me and punching the stripe onto my shoulder. Hurt like hell, but I was beaming.

Graduation day the battalion of recruits marched onto the parade deck to the accompaniment of the Parris Island Marine Band. Once we were in position, seven of us from each platoon were detached to stand in a separate formation where we would receive meritorious promotions to Private First Class. It was the custom for the Drill Instructors to select 10% of the platoon for meritorious promotion, and I, proudly, was one of them.

I can hardly express my joy at that moment - joy not of having ended the most grueling military basic training on the planet, but joy of having become a Marine. Failures of the past faded from view. I had accomplished something that only a small percentage of Americans in history had done. I had become a Marine. I was now a part of the most elite fighting force in the world.

After the graduation ceremony we were dismissed and most of us rushed to meet our families. After the hugs and kisses from Mom and Elaine, and the handshake from Dad, I had a moment alone with Elaine, and she said, "Dad was crying when they announced your promotion."

I had only seen my father cry once, and that was when Ma died. He never expressed his emotions very well.

Now it was on to Camp LeJeune, North Carolina, for infantry training.

I would be remiss if I didn't tell you about some of my fellow "Boots." They were an interesting cross-section of America. Diagonally across from my rack to the left was Willie Ryles, an ebony black man from Georgia. Willie looked like someone's statue. He was an extremely well-built guy who gained his physique from farming and football. He never said much and read his Bible every night.

My fellow Marines who were meritoriously promoted upon graduation were:

John Hurd was one of the acknowledged platoon leaders and was one of our squad leaders from day one. He was also being screwed by the Marine Corps. He had completed the Marine PLC (Platoon Leaders Course) program (a summer "Boot Camp" for prospective officers at Quantico, Virginia, that was usually completed during the summer before senior year of college). But he flunked one of his final semester courses in senior year at the University of Vermont. He was three credits shy of graduation. When he got his draft notice, he appealed to the Marine recruiter who told him that if he finished Boot Camp, they would then send him to Quantico for commissioning as a Second Lieutenant. Some recruiters lie. John never got his commission, but of the many people I met in my Marine Corps experience, he was one of two who were most deserving of commission.

Fortunately John went to school to become a Marine cartographer (map maker) and, as far as I know, never had to see any shots fired in anger.

Craig Coles was the white mirror image of Willie Ryles. Craig had three years of college and was a superb gymnast with a body to match. One day during pugil stick training, he and Willie Ryles went back-to-back and took on 3, then four, then five opponents and vanquished them all.

Herbert Malcolm Waldo Brigham III also had three years of college, had a wry sense of humor, and was a natural leader.

Angel Otero was from Puerto Rico, and while he had language and literacy issues, he was also a smart and gifted leader and a nice guy.

Curtis Brown, who travelled with me from Connecticut, was also a terrific guy and a fine Marine.

And Jimmy Cooper. By far the most interesting of my platoon mates was Jimmy Cooper. Jimmy was ebony black. He was originally from Harlem, but his mother sent him to Norfolk, Virginia, when he was about twelve to keep him away from the gangs in New York, and to be raised by his grandmother.

Jimmy was the blackest looking man I ever met. He stood about five foot ten, was lean and wiry, and could stop you cold with a look. He had been in constant trouble with the law in Norfolk. One night down on the docks, his posse got into it with a rival gang, and he took a dock hook and sliced an opponent diagonally from belt line to shoulder. He kept getting probation because his offenses usually fell into the category of "self-defense." Because he continued to do well in school and his grandmother was a well-known and respected 'church woman,' he kept getting probation. The final straw came when he was at a high stakes poker game one night in a warehouse on the docks.

The poker players were all carrying a lot of money, and as a precaution it was agreed that no one was allowed to carry a gun, but that one .45 caliber pistol would be set in the middle of the poker table in case anyone tried to come in and hold up the game.

Sure enough, somebody from a rival gang decided that the game would be easy pickings and came in to rob the players. Jimmy picked up the .45 and put six rounds in the would-be robber's chest. The judge ruled self-defense, but said, "That's it, it's either jail or the Marine Corps."

Jimmy was the most respected man in our platoon. He was not physically the strongest — there were many in the platoon who could take him in a heartbeat. He was the strongest mentally. He carried himself with a will that said, "Follow me." And we all did, gladly. He ended up being the Platoon Guide, earned meritorious promotion to Pfc., and was given the honor of being the Platoon 2017 Honor Man at graduation, for which he was awarded a set of Marine Dress Blues. He was one good man.

I remember getting on the bus to leave Parris Island for Infantry training at Camp LeJeune, and trying to stammer my thanks to my three DIs. They just shook my hand and said, "Good luck, Marine. Now get on the bus."

That afternoon they went to pick up another platoon.

2.

MARINE INFANTRY TRAINING

Right after graduation from Boot Camp, we got on military buses and headed north on coastal highway Route 17, our destination Camp LeJeune, North Carolina, along the Atlantic coast. Our home for the next six weeks or so was a subordinate camp to the main Marine Corps Base. Our new home was Infantry Training Regiment (ITR) at Camp Geiger, a 'suburb' of Camp LeJeune. At Geiger, all newly minted Marines were to complete infantry training before going to their next assignment.

Toward the end of Boot Camp, we were all given our Military Occupational Specialty (MOS) assignments, based, allegedly, on the profile we created for ourselves in the battery of aptitude tests we had taken early in recruit training.

These assignments may also have been created at the whim of a low-level government service employee working somewhere in the bowels of Headquarters Marine Corps in Washington. Some of our assignments made perfect sense, and others left us baffled.

Those of us who were going to non-infantry assignments would attend the abbreviated six-week Infantry Training Course and then move on to specialty schools, and those who were going on to infantry assignments would complete the nine-week course before being sent, in all likelihood, to Vietnam. I had wanted Infantry, but when the assignments came out at Boot Camp, I was given Supply. I was greatly disappointed, but SSgt. McDowell, an Infantry veteran of Vietnam, said, "Don't be disappointed. You just won first prize in the lottery. You'll be in the rear with the beer." Somehow I still wasn't happy. Supply was not what I signed up for.

The platoon we had bonded with at Parris Island now began to scatter.

One of the neatest things happened immediately upon our arrival at Camp Geiger. At the entrance to the camp, a Marine Corporal got on the bus and said, "Good morning, Marines." It was now official. We were no longer maggots, pukes, turds, or worse. We were Marines.

My new training company was "N" (for November) Company and consisted of about 200 newly minted Marines who were to be divided into four platoons. Prior to our being assigned to one platoon or another, we were assembled into a loose square in front of the company headquarters office. The Company Commander, First Lieutenant Garr, was a former enlisted Gunnery Sergeant. In a command voice you could hear downtown, he welcomed us and gave us a brief overview of the training we were about to experience.

Lt. Garr, I should note, was what the Marine Corps called a 'Mustang' officer. He had been an enlisted man during the Korean War. Apparently when his unit had been attacked, and the attack subsequently repulsed, (then) Corporal Garr had been wounded in one of his feet. Another of his platoon mates was more severely injured, so Garr hoisted him on his back and hiked several miles back to their base camp. He was awarded the Silver Star, the nation's third highest award for bravery.

When Vietnam broke out, he was a Gunnery sergeant. As the Marine Corps needed more and more officers, men like Garr were given the option of taking a temporary commission as an officer. It was an option he accepted. That meant that if he retired at twenty-years, he would retire initially at the highest rank he would have attained as an enlisted man.

An example of what that meant was suppose he was a Gunnery Sergeant with sixteen years under his belt when he accepted the temporary commission as a First Lieutenant. Had he remained enlisted, he would have been eligible for promotion to at least First Sergeant maybe before twenty-years and Sergeant Major at twenty-six years. If he retired at twenty years, his retired pay would be based on him having been a Gunnery Sergeant (or maybe First Sergeant). His name would have been retained on the enlisted promotion list allowing him the freedom to decide if he wanted to remain in the Officer ranks or revert back to Enlisted status. Then at the thirtieth anniversary of his original enlistment, his retirement pay would be based on the highest rank he attained as an officer. If he stayed in for the full thirty, he would retire at his highest officer rank. This program was a great win-win for the Marine Corps as well as the individual Enlisted man.

Lt. Garr was a terrific training command officer. When we arrived, he was a first Lieutenant. By the time we graduated, he was a Captain. We thought the world of him. After his welcoming overview, he introduced us to the company First Sergeant. First Sergeant Delory was impressive. He stood about six foot two or three and went about 225 pounds. He had a booming command voice that got our attention.

1st Sgt. Delory directed all the Marines who had been meritoriously promoted to Pfc. at Parris Island to step forward and form a line. Then without saying a word, he walked up and down the line stopping to look into each of our eyes for a few moments. After he had done this, he indicated that Jimmy Cooper and I, as well as several other Pfcs, should step forward. He looked at me again, asked me my name, and said, "You are the new Student Gunnery Sergeant." He assigned Jimmy to be the Student First Platoon Sergeant. After assigning others to the Student Platoon Sergeant billets of the other four platoons, he read off the remaining names for assignment in one platoon or another.

The Student Gunnery Sergeant was the First Sergeant's eyes, ears, mouth, and legs. I was responsible for relaying First Sergeant Delory's orders to the company and ensuring that the Plan of the Day (POD) was carried out. I was also to give him any and all feedback from the troops regarding training, meals, leadership issues, and the like.

I was also responsible for posting the weekly Fire Watch and Interior Guard assignments, setting up a Duty NCO Watch Plan, ensuring that an accurate count of students was maintained at all times, setting up a plan of barracks supervision for any Marines waylaid medically for any reason, and ensuring the general order of the company. All complaints and concerns came through me to the First Sergeant. I also had to complete my infantry training just like everybody else.

My first assignment was to have each of the Student Platoon Sergeants designate their respective platoon squad and fire team leaders from the remaining pool of Pfcs.

Jimmy Cooper and I became bunkmates and our rack was at the end of the First Platoon squadbay next to the entrance to the company office. The reason for that location was so I could be on-call 24 hours a day. My job was made relatively easy because Jimmy set the standard for the other platoons. He was an amazing leader and everyone did what he said. The other Student

Platoon Commanders emulated him as best they could, and routinely asked for his advice when they were uncertain about a course of action.

We had four Regular Marine sergeants who oversaw the daily training schedule – Sgt. Richard Denoyer, Sgt. Troy Neely, SSgt. Roosevelt Brown, and one other whose name escapes me. Denoyer was easy going enough, and SSgt. Brown was bigger than life, both literally and figuratively. He stood about 6 foot 6 inches and must have weighed 250 pounds, and his voice boomed all over the place. Sgt. Neely was the strange one. An infantry squad leader in Vietnam, he stood about five foot nine and was of wiry build. He had the look and demeanor of an assassin.

Assisting these Sergeants in their daily rounds were three or four Pfcs who had already finished the training and were in a "hold" status awaiting the disposition of their applications for the Enlisted Commissioning Program. They wore special armbands designating them as "Assistant Platoon Handlers," and that status, to our recently-graduated-from-Parris-Island eyes, put them in a sort of demi-god position slightly below sergeants.

One of these Pfcs stood out from the rest of the pack. His name was Tom Riedenbach, and even the four sergeants held him in a sort of awe. Tom stood about 6' 3" and had a presence about him that commanded respect. Fair haired with a solid square jaw and carrying himself with perfect military posture, he looked like the Corps was made for him. Hell, the Corps should have used him on a recruiting poster.

Tom had been in a hold status for about three months before we arrived. His situation was similar to that of John Hurd, my Boot Camp friend, who had completed the college Platoon Leaders' Course, but had to go through Boot Camp because he fell short of the credits required for graduation.

Tom Reidenbach's case was almost identical. He had completed all but a handful of credits at the University of South Carolina, having completed the PLC program. Because he was short of the required academic credits to graduate, he couldn't be commissioned in the Corps. Instead he got his draft notice. He immediately went to the Marine recruiter in his hometown of Columbia, SC. The recruiter assured him that all he had to do was complete Boot Camp and he would receive his commission through the Enlisted Commissioning Program.

Surprise! Surprise! As TV actor Gomer Pyle, USMC, would say. Some recruiters lied.

When I met Tom he was again appealing his case claiming that he had already completed the OCS requirement through the PLC program, and that he should receive his commission directly. I got to be great friends with him, and no one was more deserving of becoming a Marine Officer, nor would have made a better one.

Our training was split between the classroom and the field. We learned the art of compass reading and cross-country navigation, Fire Team and Squad tactical movements, and how to use just about every weapon in the Marine Infantryman's arsenal – not just the M-14 rifle, but also the grenade launcher, the M-60 machine gun, the 12 gauge pump shotgun, the .45 caliber pistol, the new Light Assault Anti-Tank weapon (LAAW), the older 3.5mm rocket launcher, the .50 caliber "Tommy" gun, the newly introduced to the Marine Corps M-16 rifle, as well as how to properly pull the pin and throw a live hand grenade.

Administratively I began to develop an appreciation for the challenges of keeping all the Marines in the field at the same time. There were always training casualties – sprained ankles, pulled muscles and the like, and as the weather got colder (November through the end of February in North Carolina was generally cold, damp, and raw), we saw the onset of ARD – Advance Respiratory Disorder.

The barracks we lived in were built during World War II. They were one-story buildings made of cinder blocks. They were heated by a forced hot air system. This system did keep the barracks warm, but the dry forced hot air, loaded with dust, was a shock to the respiratory system that had been running around all day in the boondocks sucking in cold, damp air. The resulting dust mites kicked up by this archaic heating system played havoc with the lungs of Marines in training. We suffered a number of ARD casualties hospitalized. Then we had to deal with the returned casualties from the hospital who were now placed on "light duty." These Marines in the process of final recovery from ARD required at least a Pfc. to be in charge back in the barracks while the rest of us were completing training, so if there was no Pfc. on light duty, I had to assign one to stay back in a supervisory role.

Near the end of training, we almost lost Jimmy Cooper. In the last week at Boot Camp at Parris Island, Jimmy had been hospitalized in the middle of the night with a severe intestinal disorder. It was something terribly painful, not appendicitis, and not otherwise conclusively diagnosed by the physicians.

He apparently had some kind of intestinal blockage. After two nights in the hospital at Parris Island, Jimmy was told that a third night would require that he be set back to complete recruit training with another platoon. He insisted that he was better, got out of the hospital, and rejoined us for graduation. In retrospect, his pain must have been excruciating, because the condition hadn't gone away. But he had such an enormous amount of self-discipline, he powered his way through it and was named our platoon honor man.

Whatever his ailment was, it came back in spades while we were entering the last week of training at ITR. He was rushed to the hospital, and upon initial diagnosis, the American Red Cross was notified to contact his grandmother with the recommendation, because of his dire condition, that she come from Norfolk, VA, to Camp LeJeune.

Jimmy was in really serious medical trouble. They finally operated on him and found a blockage of some kind in his intestine. The First Sergeant and I went to visit him in the hospital, and Jimmy was in pretty weak condition. His grandmother was right out of Central Casting: a sweet, devout, Black Christian lady who insisted that her prayers would get Jimmy through, and about a week later he was back on limited duty.

She might have been right.

And our training continued.

Our lives outside the training regimen were limited to the confines of Camp Geiger. None of us had a car, nor would we be allowed one until we had completed the training. We regularly went to the Post Exchange (PX) to replenish supplies and renew our taste for junk food, which had been forbidden at Parris Island. I can't tell you, based on the months'-long deprivation of our taste buds, how good the cheeseburgers tasted at the Enlisted Club. By today's standards, they were probably crap. But Marines have never been accused of being fussy when it comes to food.

Our free time in the evenings and weekends, when not standing one kind of duty or another, was spent spit-shining our boots, polishing our brass, and maintaining our uniforms. We took advantage of one of the commercial laundry services whose trucks showed up in our parking lot every evening. They would pick up our dirty uniforms and deliver them back to us the next day with everything but the underwear and socks starched and pressed.

(A tragic coda to this experience was when in recent years it was discovered that the most contaminated ground and well water anywhere in the country was at Camp Lejeune. And the pollution was significantly contributed by the dry cleaning companies that daily laundered and starched the uniforms of probably over twenty-thousand Marines every day. We Marines in our starched, pressed utility uniforms, looked more squared away than any other service, but there is, in fact, a Law of Unintended Consequences.)

A fond memory was of the many evenings we would spend sitting on our footlockers talking and lying to each other about joys we experienced in civilian life while we spit-shined our boots. We were a group of guys who were very content. We were Marines learning more and more about the skillset required to become a warrior. During those evenings after training and chow, it was a pleasure to sit and talk while we shined boots, polished brass, and did a myriad of other tasks to ensure that we were sharply 'squared away' when we turned out for duty.

During the second week we were at Camp Geiger, my routine was interrupted one day when the First Sergeant bellowed for me to get into the office immediately. I promptly went in and said, "What's the matter, First Sergeant?" He said, "Why didn't you say that you were eligible for Officer Candidate School?'"

I replied that as far as I knew I was not eligible, and then related what I knew from my experience in Boot Camp. He said that the letter I got was incorrect. It turned out that if you got one score below the 35th percentile, but the rest of your scores averaged much higher (I forget the number), you were eligible. It turned out that my only 'below average' score was 34 point something percentile. Averaged in with my other four scores my composite score was well above the minimum required for acceptance into the Marine Corps Enlisted Commissioning Program. So First Sgt. Delory asked me if I was interested in becoming an officer. I had to think that one over. At Boot Camp I thought it was a terrific idea, but after having received a rejection letter, I was now thinking that maybe I should just do my two years and get out. Becoming an officer meant entering a new three-year contract.

I asked how much Second Lieutenants were being paid, and the First Sergeant said, "$415 per month." Well I was making $112 per month as a Private First Class, and I was planning on getting married, so I said, "Sounds like a good idea to me."

There were several stages to my being accepted into the program. The first step in the screening process was the interview with the November

Company Commander. I met one-on-one with now Captain Garr. The CO stood only about 5' 6" but carried himself like a much taller man. He was impressive. He also was very friendly when I reported to Him. "So you're thinking of becoming a Marine Officer, are you?"

"Yes, Sir, I would like that very much."

"Outstanding," he said. "Now here's what you have to go through to get from here to there." And he laid out the hoops I was going to have to jump through just to get accepted into the program, and then what I would go through at Officer Candidate School (OCS). I told him that I was ready to get the proceedings started, and he had the First Sergeant bring in all the paperwork.

The paperwork was pretty comprehensive. I had to list every place I had lived; every employer I had ever worked for; and who my friends and neighbors were. I also had to send away for my high school and college transcripts (that part worried me) and get written recommendations from several people.

After a couple of weeks, I had the full package prepared and submitted it up the chain of command, through the company CO to the battalion commander to the Regimental commander to Headquarters Marine Corps, etc. The package passed the initial screening, and I was scheduled for a series of interviews, first with a team of company commanders, and, if I got through that one, the next would be a battalion screening board. Finally I would have to have a personal meeting with the regimental commander, and if that went well, the package would be submitted to Headquarters Marine Corps (HQMC) for final disposition.

I, of course, was a bit nervous about these screening boards. First I met with four training company CO's, peers of now newly promoted Captain Garr. I prepared like I had prepared for final inspection at Parris Island. I had the most squared away uniform and mirror-gloss, spit shined shoes as I reported to the battalion headquarters conference room. The interview lasted about an hour and went very well. Capt. Garr and First Sergeant Delory had coached me on my demeanor and how to respond to questions. For example, if I was asked how well I liked a particular part of the training and I didn't like it, I was to be honest but respectful in my answer.

The company COs forwarded my application "With enthusiasm" to the next level – the battalion screening board. This board was made up of three majors and two senior captains, all of who were on the battalion staff. One of the majors was the Executive Officer (XO), the number two officer in the battalion.

I had gained some confidence from my experience with the previous board, so once again I made myself look as sharp as possible and reported "... to the battalion screening board, as ordered, Sirs!."

This one was the real "stress" interview. These guys were good. We sat at a round table in a conference room and I began to feel as though I was a basketball being whipped around by a circle of Harlem Globetrotters. I would field a question from one officer to my right, maintaining good eye contact with him with the occasional look at the other officers, then from the left end of the table I'd get another question. After answering dozens of questions on the training at Infantry Training Regiment, as well as other Marine Corps related questions, and also the obligatory 'Why do you want to be a Marine Officer?," the XO asked, "Do you smoke?"

After all the other routine batting practice speed questions, this one was like a 98 mile per hour fast ball headed right between my eyes and delivered by a crafty pitcher.

How do you answer this question? What does the questioner want?

Sucking up all the courage I had, I looked him right in the eye and said, "Yes, Sir, I do." He then said, "What do you think of the Surgeon General's new warning on cigarette packages?" (In 1968 the Surgeon General of the US had issued the unprecedented dire warning that was required to be posted on every cigarette package.)

I thought for a moment and said, "Sir, at Boot Camp we were allowed three cigarettes a day. When the Drill Instructor would call out, "SMOKERS DRAW ONE," we would go to our foot lockers, get one cigarette and a pack of matches, then on the DI's order we ran outside and formed a 'smoking circle. One of our squad leaders stood in the center of the circle with a pail filled with about two inches of water. At that point, we would all shout at the top of our lungs, "SIR, THE SMOKING CIRCLE FOR PLATOON 2017 IS NOW FORMED, SIR. EVEN THOUGH THE SURGEON GENERAL HAS WARNED US ABOUT THE DANGERS, WE REQUEST PERMISSION TO LIGHT UP BECAUSE MARINES USUALLY DON'T LIVE LONG ENOUGH TO GET LUNG CANCER, SIR!"

"Does this answer your question, Sir?"

I got a hearty laugh from all of them, and they endorsed my application "With enthusiasm" noting, also, that Pfc. Power had, under duress, ". . . exhibited an evident sense of humor."

I got that piece of classified information from my friend Pfc. Tom Reidenbach who managed to charm someone at Battalion HQ to give him a copy of the evaluation.

Next stop, the regimental commander.

Here I was a newly minted Marine who at Boot Camp had never seen my First Lieutenant platoon commander close up enough to even say, "Good morning, Sir," after the first day of training when he gave his initial class on the Uniform Code of Military Justice. Frankly, I couldn't have picked him out of a police lineup. Now I was on my way to have a personal meeting with a full Colonel.

Actually the meeting went very well. The Colonel put me right at ease. We sat at the coffee table in his office, me sitting forward on the edge of his leather couch, and himself in an easy chair. He actually did most of the talking, telling me about the rigors of OCS and Basic School (the 26 week course following OCS). We had a pleasant meeting, and he, too, forwarded my application onward "with enthusiasm" to its eventual resting place at HQMC.

It was now mid-December 1968, and our training was winding down to its conclusion. At the end of training, the troops were dispersed in different directions to various schools where they had been assigned. I was to have gone on to Supply School, but because of my OCS application, I was now put on what was called an "Administrative Hold." I would stay on at November Company as a training assistant until HQMC made a final decision on my OCS application.

Jimmy Cooper was to have gone on to Cartography School to learn how to make maps (which he eventually did do), but because of his illness and operation he was in a "Light Duty" status for a month. That meant that he could not participate in physical training until he got clearance from the doctors. He had a scar from his navel to his groin from the surgery, and it matched the scar he had from his navel to his sternum from a pre-Marine Corps knife fight back in Norfolk. He was moved over to "Casual Company" where assorted Marines are assigned for various reasons. Some might just be awaiting a regular discharge, some might be recovering from an illness or injury prior to their next assignment, and some were awaiting judicial punishment of one kind or another.

Jimmy, assigned to one of the "good guy" units, was given an assignment as a "Brig Chaser," that is, he was tasked with bringing back Marines who had gone on "Unauthorized Absence (UA)" – that was for less than 30 days – or "Absent Without Leave (AWOL)" – which meant gone for more than 30 days. Said Marines were usually in the local custody of some municipality, or, if they had been 'over-the-hill' for a much longer time, they might be in federal custody.

It was Jimmy's task to go to the appropriate authorities, retrieve these errant wanderers, and return them to Camp LeJeune where they would receive an appropriate form of justice.

He told me about his briefing from the Provost Marshall (PMO), a Marine Major, before going on his first assignment. The PMO started issuing Billy clubs to those assigned as chasers, and when the PMO got to him, Jimmy said in his classic authoritative voice, "I don't need no stick." The PMO just looked in Jimmy's eyes, and said, "All right."

And Jimmy never did "need no stick." I think all the people he brought back from these various forms of unauthorized absence intuitively understood that with Pfc. James E. Cooper, you never questioned his authority. And you didn't want to be stupid enough to run from him. Whatever was waiting for them at Camp LeJeune was better than crossing Jimmy Cooper.

Meanwhile I was now an Assistant Troop Handler with November Company, and I was living in an area of the barracks blocked off from the trainees by a line of wall lockers. I shared the space with Tom Reidenbach and three or four others who were awaiting orders to OCS. I oversaw and mentored the work of the new Student Gunnery Sergeant for the next training group, as well as did the bidding of the platoon sergeants who managed the daily training schedule.

With the Christmas holidays approaching, First Sgt. Delory called me in and said, "There is going to be a competition for the best Christmas decorations, and right now we don't have any." I responded, "I'm right on it, First Sergeant."

I called a meeting of the Pfc. billet holders for the new training company and told them we needed some sort of Christmas display, and I expected them to come up with something.

The next morning the First Sergeant and Captain Garr arrived at our company headquarters to find that a crèche had been built (out of stolen lumber), and inside the crèche (bordered with North Carolina pine boughs) were St. Joseph, the Virgin Mary, the baby Jesus, and a couple of shepherds, all represented by stolen silhouette targets from the rifle range.

We didn't win the contest, but I really have to hand it to Marines for innovation and originality.

Actually the Training Assistant job was kind of boring. We TAs would go to the field with the troops, but mostly we stood on the sidelines while they went through the training. Our main job was to do what the sergeants weren't interested in doing.

But Christmas was coming, and I had requested leave for the holidays. I was granted leave from December 26 through January 5 and was looking forward to getting home. We sent the newly acquired training company home on leave around December 23, as did all the other commands, and the base was really empty. Still, the work and requirements of the Marine Corps went on, so those of us who remained had to stand duty.

I was assigned to stand as the Duty NCO at battalion headquarters from 4:00 PM December 24 through 4:00 PM Christmas Day. I dutifully reported to the battalion XO on Christmas Eve and he gave me the logbook and all appropriate instructions. Within a half hour the headquarters building was empty. Fortunately I brought a book with me (I think it was Michener's "Hawaii") or I would have died of boredom. The Staff Duty NCO, a Staff Sergeant, stopped by early in the evening, giving me enough time to get to the mess hall to eat dinner, but after I got back to battalion headquarters, that was the last I saw of him. I never did see the Officer of the Day.

I locked the place down around 10:00 (2200 hours) and went to sleep on the duty rack that had been provided. Got up around 0600 Christmas Day and waited for someone to come and relieve me so I could go to breakfast (actually holiday brunch – on Sundays and holidays the mess halls served a brunch and a late afternoon dinner). Unfortunately no one showed up. Nor did anyone show up in the afternoon so I could get Christmas dinner. Finally around 3:30 the Battalion Commander showed up. I presented him with the log book in which I had posted only a few entries – basically "0600 – nothing to report." But when he saw that I had recorded "No relief for morning chow" and "No relief for afternoon dinner," he hit the roof. He said, "You mean to tell me that no one has been here since yesterday evening?" I said, "Yes, Sir." He then said for me to get down to the mess hall and get dinner and he would relieve me himself. Wow. A Lieutenant Colonel doing a Pfc's job. *That* taught me so much about Marine Corps leadership, and the lesson served me well in the future.

I guess some heads rolled on that one. When I got back from Christmas leave I found out that there were a couple of reprimands issued. The Marine Corps is obsessive about leadership, and the fundamental objectives of leadership, as taught by the Marines, are "Accomplish the mission, and take care of the people." When you look at these objectives, it is obvious that you can't accomplish one without accomplishing the other. And when a senior officer finds out that junior officers and Staff NCOs are not taking care of their people, then it's going to be someone's ass.

In any event, the day after Christmas I got a ride to Connecticut from Sgt. Denoyer and his wife who were going to their hometown near Boston for the holidays.

It was good to be home. My family in Norwalk, CT delayed Christmas so we could celebrate together. I also got a chance to go around and visit with my friends, most of whom were still baffled that I volunteered for the Marines, of all things. But it was a good break. I also used the opportunity to trade in my 1966 Austin Healy 3000 (a huge mistake) on a 1967 Mustang fastback with a 289 cubic inch V-8 engine, 4-speed floor transmission, and glass-pack mufflers for that really mellow sound. I was concerned that at Camp LeJeune and its surrounding environs, I might not be able to find mechanical service capable of caring for a relatively rare British sports car that required an engine tune at least every thousand miles. So to be safe, I traded it for the Mustang. Don't get me wrong, I loved the Mustang, I loved its deep throated purr, and it was one slick looking machine. But the Healy was my treasured baby.

(An aside note in late December 2020, there is a company that is reproducing 1967 Mustangs – the Shelby version – for $120,000 per copy. This is interesting as the original 1967 Mustang sold for a little over $2,000.)

New Year's Eve my parents allowed me to have a party for my friends as well as some cousins who came down from Waterbury. Kathy came up to join us and we had a great time.

I drove the Mustang back to LeJeune where I had permission to keep it now that I had completed training. It proved indispensable for driving sergeants where they wanted to go. I was the equivalent of today's Lyft or Uber services. I ferried sergeants everywhere. And with no tips.

One night in early January it was particularly cold. It was bitter cold with temperatures dropping down below freezing, a rarity in coastal North Carolina. We were slipping into the 'teens with a serious wind blowing. In fact, the wind was howling. Just before we climbed into our sleeping bags for the night, Staff Sergeant Roosevelt Brown bellowed out, "The Hawk (wind) is comin' to get us all!" We were not reassured by his weather prediction, but as we observed him, he appeared, in his massiveness, to be impervious to the whims of mere weather.

I stripped down and crawled into my sleeping bag, bringing my boots and uniform in with me. Nothing like putting your feet into warm boots on a cold morning. I zipped the sleeping bag up to my neck and lay there looking into the dark and frigid North Carolina night. Someone later told us that the temperature dropped to 15 degrees overnight – a rare occurrence for North Carolina. All I knew was that I was nice and toasty next to the fire and looking up at the stars.

I should describe the sleeping bags for you. They were down-stuffed marvels that enveloped us and kept us warm under the most severe circumstances. They were designed so that your whole body, including your head was encased – minus, of course, your face – and protected from the outside elements. When setting up one's field sleeping accommodations, one first hollowed out a bed in the sandy North Carolina soil. Next was to fill the hollowed out area with pine needles (in amazing abundance in North Carolina) for both cushioning and air circulation. Next, lay down one's poncho (a rubberized rain protector), and then put down the down-filled sleeping bag. The Marine Corps insisted that this was the equivalent of sleeping on a Serta mattress.

All one had to do was climb in, zip up, and go to sleep.

So there I lay, exhausted from another long day, letting the dying fire keep me warm while my own body heated up the inside of the sleeping bag. The stars were bright and plentiful as I lay there, content in my minimalist Marine coziness. I saw shooting stars and heard the drone of the training sergeants' fireside conversations fading into the background, much as when as a child riding in the back seat I heard the conversations my parents fading away, and then blissful sleep.

Until about sometime around zero dark thirty when I had turned around inside my sleeping bag. At that time of my life, I always then slept on my stomach, but on this night I awoke in absolute blackness with the cloth of the sleeping bag pressed into my face.

I moved immediately into panic mode as my heretofore undiagnosed claustrophobia moved into red-line territory. I totally freaked out and was thrashing around in my sleeping bag trying desperately to find the opening for my face. But those damned bags were snug. I finally managed to free an arm and in a desperate grab, pulled the face opening around so I could see and breathe normally.

As I lay there gasping for breath, I looked up to see the massive presence of First Sergeant Delory standing over me. He looked down and said, "Power, just what the fuck is wrong with you?" I realized that I must have looked like a totally deranged worm.

Somehow I moved past that, but the incident was to repeat itself nearly ten years later when I was a Captain and training my Marines in the fields of Pennsylvania.

In mid-January 1969 I was notified that I had been accepted to OCS for the class beginning February 10 of that year, and that my orders would be forthcoming. I felt badly for Tom Reidenbach, as his appeal had been turned down. Rather than go through OCS, a repeat for him of the Platoon Leaders Course, he elected to remain an enlisted man and go instead to Avionics School to become certified to do electronic repair and service work on fighter jets. By the time I saw Tom again in 1971, he had been meritoriously promoted to Lance Corporal, Corporal, and Sergeant. That was no surprise.

My orders finally came through and I headed north in late January to spend a few days at home before reporting to OCS on February 10. I also made arrangements for Kathy to take the Mustang while I was in OCS as she would be doing her student teaching in Baltimore during the same period and would be commuting back to her home in Arlington for the weekends. The plan was for her to go home Friday nights and then come down to pick me up on Saturday afternoons and bring me back to Arlington to stay at her house for weekend breaks.

3.

OCS AND BASIC SCHOOL

Abraham Lincoln once told this story to illustrate some point to his Cabinet: "I once heard of a man who was about to be tarred and feathered and run out of town on a rail. The man looked at his tormentors and said, 'If it weren't for the honor of the whole thing, I'd just as soon walk.'"

Since I was about 12 years of age, I had read books about the military – from the Revolutionary War through Vietnam. I was always impressed with the Marines, and when I started college at Mount St. Mary's, I planned to be an Officer of Marines. Back at that time I signed up for the Platoon Leaders Course with the Officer Selection Officer who visited the Mount every few weeks. Unfortunately, I shot myself in the foot academically and temporarily had to give up on my plan until, conveniently, the opportunity presented itself again during my enlisted training.

Now here I was, Pfc. Brian Power, One Each, Green (in the parlance of the Corps), in mid-January 1969, being granted terminal leave from the Infantry Training Regiment (ITR) at Camp Geiger, Camp LeJeune, NC, to detach and report to Officer Candidate School (OCS), Quantico, VA, reporting no later than 1600 hours on 10 February 1969.

I remember when I got the word that my orders to OCS came in, I was in the field assisting the NCOs with a new company of trainees. We had been out a couple of days and I was short on sleep and cleanliness. When the word was radioed out to me to return to the company area, I left the field and drove back to our company headquarters. After talking with the First Sergeant, I went about the process of checking out with all the appropriate areas (Supply, Armory, etc.), as quickly as time would allow, and said farewell to my friends. I went back to my barracks, showered and packed, checked out with my Commanding Officer and again with the First Sergeant, and got into the '67 Mustang pointing the hood North toward Washington, DC.

The evening I left Camp Geiger and headed north to a better future, I was simply anxious to get out of there. I should have waited an extra night to get catch up on my sleep and depart the following morning. But in my haste to get to the future, among other things, I had neglected to service the Mustang properly. Leaving the base on a late mid-January afternoon, it got dark early. I was in the middle of a two-lane road in East Jesus North Carolina when, a couple of hours later, the Mustang began to buck. As it turned out, all those trips to the field ended up having an impact on my poor car.

Being a bit short of sleep, this turned out to not be such a bad thing. I was nodding off when the car lurched, continued along, then lurched again. I managed to nurse the car to an open gas station in the rural boonies where a kindly man, after looking at the Marine Corps gate pass sticker on my windshield said, "Ah know what yer problem is."

Having observed not just the windshield sticker, but also my haircut, he said something like, "Dust." He flipped up the hood, unscrewed the air filter above the carburetor, took the filter cover off, turned it over and whacked in on the ground a few times to produce a cloud of dust, and said, "Y'all cain't go drivin' off into the boonies without pickin' up a bunch of dust. Y'all be more careful now, y'hear?"

I heard. And thanked him and paid for his services. All that ferrying of NCOs back and forth from garrison to the field in the dusty boonies of North Carolina had filled my Mustang's air filter with dust. All things considered, it kept me alive as it woke me up when the car started bucking.

I remember one more thing about that drive. The overall drive from LeJeune to DC was just over five hours. Another couple of hours after my emergency pit stop, it was approaching 11:00 PM. Totally exhausted from my field experiences, as I was coming up Route 95 North and passing through the towns just north of Quantico and just below the Washington Beltway, my tiredness really started catching up with me. By that time, I had been on the road for at least five hours with one break to fix the car, and another break to get a burger, fries and shake. For two-days I had not had a lot of sleep. I began hallucinating that 18-wheel semi-tractor trailer trucks were nothing more than swaying elephants in my way. Between opening the windows all the way and playing the radio as loud as I could, I managed to snap out of my stupidity long enough to get to Georgetown University in Washington, DC.

I spent a night with my old high school friend Dan Sullivan at his dorm at Georgetown. With the anti-war fervor on college campuses at a fever pitch, some of his dorm-mates were barely polite in their introductions to me. I can't say they were rude. But I remember I could feel their discomfort. Nonetheless, they respected Dan, who was in Army ROTC and due to receive his commission as a 2nd Lieutenant upon his graduation in May. He held their denunciations of the military to a minimum. I am sure that they thought that as a Marine I had just completed serial killer training. It would not be the last time I got that impression from people who just didn't know. That was how the public perceived the military in 1969.

Some brief chat, a couple of beers, and I was out for the night. The next day, having caught up on some needed sleep, I thanked Dan for his hospitality and drove home to Connecticut to spend a few days with my family. It was great to see everyone. Coming home kept me in touch with the real world, particularly as the Vietnam War was churning on.

After a week or so, it was time to move on to the next phase of my life. I packed up, said goodbye to Mom, Dad, Elaine, and David, and headed south. Five hours later I was in Arlington, VA.

I spent the final night before reporting to OCS at Kathy's home in Arlington. On a sunny, cold, clear and crisp winter day, I remember having breakfast with Kathy's mom Anne. Kathy was, at that point, completing her final semester at St. Joseph's College in Emmitsburg, MD. In the next couple of weeks, she would be starting her six-week student teaching assignment at a grammar school in Baltimore. Over coffee her mother and I had a great chat in the den, and, as a fully experienced military wife, she was totally encouraging as I gave her a hug and headed for my new assignment. I drove thirty miles south to Marine Corps Base Quantico and checked in to OCS on that sunny, cold morning of February 10, 1969.

OCS at that time was another form of Boot Camp, but there were some significant differences. Boot Camp, to me, while a physical challenge beyond belief, was tougher psychologically. In my opinion, OCS was tougher physically. Don't get me wrong – both were physically and mentally challenging. But I was fortunate that it was Boot Camp that had toughened me mentally for what I was about to experience.

We were now to refer to ourselves as "Candidate" and address our platoon sergeant as "Platoon Sergeant," and his two assistants each as "Sergeant Instructor." Again, as in Boot Camp, we had to refer to ourselves in the third person, as in, "Platoon Sergeant, the Candidate requests permission to speak."

Our platoon, one of three, was in "E", or "Echo" Company. The 40-man platoon was made up of roughly two-thirds college graduates and one-third former enlisted like me.

Our platoon sergeant was Staff Sergeant Howard who, as he claimed, was the greatest warrior from upper east Tennessee since Sergeant York (Army Sergeant Alvin York, awarded the Medal of Honor in WWI). SSgt. Howard was a real piece of work, and a fine Marine and leader. He was another shining example of superior leadership that I had both experienced and come to expect from the Noncommissioned Officer (NCO) and Staff Noncommissioned Officer (SNCO) ranks. Assisting him and each bearing the title of Sergeant Instructor were Sgt. Schuler and Sgt. Cote. Both were terrific leaders as well. Both had a huge impact on me as you will see later.

Our platoon leader was Lieutenant Matulich, a wounded (knee) combat veteran of Vietnam who was an overwhelmingly bad choice for the job. More on him later.

Among my fellow officer candidates I quickly made friends with the former enlisted men. Steve Pfeif, Jim Forney (a future Colonel), Dave Bice (who would rise to the rank of Lieutenant General – 3 stars), Wayman Bishop (a future Colonel), David Ross and Vance Page. I also connected with John Mountain, John Steele, Richard Poore, JD Richardson, Sam Gauny, Mike Reynolds, and a few others who signed on right after college. We were a pretty compatible group.

After drawing our initial issue of gear, done in a similar fashion to the way it was done at Parris Island, we settled into the training routine. SSgt. Howard informed the college grads that they were to take their lead from the former enlisted men, as we knew the ropes. Simply put, when in doubt, they were to come to one of us for advice.

Our training week ran from early Monday morning through noon Saturday. First thing every morning we fell out behind the barracks on the parade deck for morning Physical Training (PT). Let me tell you that there is nothing to compare with the raw cold wind of northern Virginia in February at 5:30 the morning, sweeping across the parade deck, as we stood there in PT gear next to a tributary of the Potomac River.

We'd be standing there in sweatshirt, PT shorts and boots, and the lacerating wind coming across the Chopawamset (sp?) Creek would cut right through us to the bones. The typical early morning workout consisted of jumping jacks, push-ups, and squat thrusts. Then we'd go back to the barracks, shower and change into our utility uniforms, clean the barracks, and head for chow.

The mess hall that first few weeks was a disaster. The food was horrible, and it seemed that every evening we had hot dogs, cold-cuts, or unidentifiable cuts of meat that had the distinction of getting larger and tougher the more you chewed. After numerous complaints from our Platoon Sergeants and Sergeant Instructors, the commander of OCS, Colonel Ing, conducted an investigation and discovered that the Gunnery Sergeant in charge of the mess hall was selling our steaks, chicken, and fish on the outside market and feeding us crap. He was promptly relieved and Court Martialed and ended up doing time in Portsmouth Naval Prison. Our food improved greatly from that point on. I don't think his did.

Monday morning through Friday noon our training day was a mix of classroom learning in the basics – Marine Corps History, First Aid, weapons familiarization, the writing of tactical orders (called five-paragraph orders), and outside exercises in close order drill, spiced daily with isometric exercises, a running of the obstacle course, a three-mile run, pugil stick fighting (more on that later), and hand-to-hand combat.

The runs were challenging for many of the college graduates. The rest of us former enlisted knew how to put it into cruise control and finish with the platoon. Some of the college guys were in really terrific shape and could run anyone into the ground, but many of them arrived in not great physical condition, and they (or the officers who recruited them) should have known better. Within a few weeks most of the slowpokes were able to keep up with the platoon, but there always seemed to be the same few stragglers. The one candidate who never could seem to keep up was Candidate C. This guy was one of three in our platoon who had Masters degrees in esoterica. C would always drop back from the platoon and dramatically collapse crying that he just couldn't make it.

He also screwed up everything else he did.

Each week we were evaluated by our Platoon Sergeant and Sergeant Instructors. Among their rating tools were "chits" they would hand out for various infractions – an improperly made rack, tarnished brass, un-shined boots, discrepancies on equipment inspections, etc. Each chit had a name – rack chit, shoe polish chit, brass chit, shaving chit, etc. Those chits counted as "misdemeanors," but if a pattern of accumulating the same type of chit developed, the candidate could run the risk of being dropped from the program. More serious were chits handed out because of a dropped rifle, or even worse, a dirty rifle. The worst kind of chit you could get was an integrity chit. Get just one of these for lying, cheating, stealing, or deliberately violating any of the rules and regulations, and you'd find yourself being thrown out unceremoniously and relegated to completing your contract as an enlisted man with a reputation for dishonesty.

Integrity. Honor and Integrity. We were hammered with it. The heart and soul of a Marine officer was his honor and his integrity. Those core concepts must be branded on our souls. I knew from everything I had read before enlisting that this was the essence of Marine leadership.

To this day I believe in that with my soul.

What happened in OCS, and Basic School as I will relate in the following narrative, would challenge a lot of assumptions I had about Marine officers.

At any time during OCS, a candidate could "DOR" – Drop on Request. If you were a college grad that meant leaving Quantico, going through Parris Island, and doing three years as an enlisted man. For the former enlisted, it meant simply going back to your last command and resuming your status in the enlisted ranks.

At the end of the fifth week of OCS (and also in the ninth week), we conducted "peer evaluations" and rated our fellow candidates in a broad range of categories. If you fell at the bottom of the subsequent rankings, you could be subject to being kicked out of the program. The peer evals, plus the ratings from our platoon sergeant and sergeant instructors, determined whether we would stay in the program or not.

When I think about it, this was basically a 360° personnel evaluation that became *de rigeur* years later, thanks to some nitwit Human Resources trained clown who hung out his or her shingle as a consultant and said, "This is the latest thing."

The Marines were doing it from the gitgo.

Back to training. Friday mornings usually meant more class and drill time, and then, immediately after noon chow, we would don full field packs with rifles, web gear, and two full canteens and head out to the Hill Trail for a cross-country hike. Cross-country hike is a term that, to a civilian, evokes romantic images of strolling through rolling fields of wildflowers with, perhaps, a light LL Bean pack on your back with a couple of sandwiches, a corkscrew, and a bottle of Chardonnay inside.

In Quantico, northern Virginia, February 1969, cross-country hike was mostly uphill. And I mean steeply uphill. Usually wet, God-awful slippery, muddy uphill. The terrain at Quantico, Virginia, is quite severe. The hills aren't necessarily all that high, but they are steep and heavily wooded. In the winter and spring, they are wet and slippery.

Our company commander would give the word and we would step out smartly. The location of our OCS company was at the OCS Annex; we were actually a mile removed from mainside OCS and the beginning of the Hill Trail. We first had to walk a mile along "Shin Splint Highway," the blacktop road that paralleled the Air Station and that connected us to mainside. Shin Splint Highway got its name because when you marched along its hard surface in combat boots, one of the by-products of setting your heels down too hard was that you would get shin splints. If you have never had them, imagine what it would feel like if someone hit you on both shinbones with a baseball bat. That's shin splints.

The first Friday we were there we took a three-mile hike over the Hill Trail (plus the Shin-Splint Highway mile going and coming). Each week they added a mile or two to "The Trail" until, by the time we reached week ten (the final week) we were going about eight thousand miles. (All right, maybe not that much.) All of this, of course, was conducted wearing a sixty-pound pack, web gear weighted with two full canteens, ammunition packs, and carrying a 9.5 pound rifle. (Today Marine Officers train with heavier loads, and I don't know how they do it.)

At the end of the second week Hill Trail trek, I had managed to pull muscles in both my right calf and my left quadriceps. The pain was a 9 out of 10, but I still managed to stay with the platoon and finish the hike. When we got back to our barracks, our Corpsman (the enlisted Navy medical specialist assigned to every Marine unit) wrote out a chit to get me into the rehab whirlpool at the mainside clinic the following morning. Saturday morning SSgt. Howard excused me from inspection and had me report to the company office to get transportation over to mainside. I hobbled over to the company office and presented my medical excuse. But the company Gunnery Sergeant (Gunny) basically said 'Tough luck, Candidate, we don't have any transportation for you, so you'll have to walk.'

(Now bear with me. This is the beginning of a tormenting experience that challenged every assumption I had about Marine officers.)

Not having high expectations of a chauffer service, and being a Parris Island graduate Marine who was conditioned to accomplish his mission, I began my hobble to mainside. I had gotten just about 100 yards down Shin Splint Highway when a white Volkswagen Beetle pulled up next to me and stopped. The driver leaned over to roll down the passenger side window and said, "Where are you going, Candidate?" I saw the shiny eagles on his collar and immediately came to attention and snapped off a salute. "To the mainside clinic for therapy, Sir." He said to hop in and that he would give me a ride. He introduced himself as Colonel Ing, the Commanding Officer of OCS, and asked how I had injured myself. I explained about the Hill Trail and he asked why the company didn't provide transportation. I said that there apparently was none available, and he said, not to worry about it, he would have a vehicle pick me up for the return trip. I thanked him very much, got out at the clinic, saluted him, and didn't think much more about the incident.

It was to come back and haunt me -- big time.

When I finished my session in the whirlpool, a Lance Corporal came in and asked if I was Candidate Power. I told him I was, and he said that he had a Jeep waiting for me and would drive me back to the company when I was ready. I finished my session, dried off, got back into uniform and met him in front of the clinic. After he delivered me to the Company area, I reported to the Company Gunny. He chewed my ass out and asked why I told Colonel Ing the company wouldn't provide transportation. I said, "Because that's what you told me." He told me to get the hell out of there and not come back for any reason. But there was more to come later.

* * *

Weekly, when we got back from our Hill Trail excursions, we'd turn-to and clean our rifles and gear in preparation for Saturday morning's inspections. Friday nights after evening chow, we were given 'Base Liberty' and allowed to go into 'Quantico Town' and get a beer or two. Saturday mornings we always had inspections.

The inspections could be on our field equipment, rifles, bunks, uniforms, personal gear in our footlockers and wall lockers, or on the barracks itself for cleanliness. Automatically included in any inspection was an examination of our personal presentation – uniforms, shave, haircut, shined boots, etc.

On this particular Saturday morning inspection, I think Ross and I flirted with being kicked out of OCS. Both of us being former enlisted, we'd be sent back to that status and have to put up with it.

My partner in crime was David Ross, future CH-46 helicopter pilot (and A-4 fixed wing pilot and C-130 Transport pilot – a rare tri-fecta), and maybe stand-up comic somewhere down the road. He and I made a terrible combination because, as fellow Irishmen, we always saw the humor in everything. I'm sure it is unintentional on the part of the Corps, but both Boot Camp at Parris Island, for all of its horrors, and Officer Candidate School at Quantico, routinely produce moments of unbelievable absurdity.

And once one of us started laughing, the other soon followed.

Okay, let me explain the latest episodes that I was convinced would result in our being sent away from Quantico because we lacked proper gravitas. We would be doomed to finish our tours elsewhere in the Corps (i.e., Vietnam) as enlisted men.

Every training platoon has a certain number of people who screw up constantly. It's just like any other profession. Of forty-two Officer Candidates, three will dominate the top of the standings both academically and in peer review. Three, likewise, will dominate the bottom of the standings. The rest of us fall somewhere in the middle.

Our platoon had three candidates with an extraordinary talent for screwing up. All three of these guys had Master's degrees in esoterica like medieval French Literature, and Sixteenth Century Florentine Art, or the Impact of the Minuet on Popular Music of the 17th Century. You get the idea. They were smart as all get out but lacked even the barest minimum of common sense. I have no idea what drove such geniuses to join the Marines. Of these three, Candidate "Sloan" (an alias) was a special case all unto himself.

OCS operated on a system of 'Chits.' These Chits were very much like the demerit systems at the US Military Academy at West Point and the Naval Academy at Annapolis. If your rack (bed) was not made properly, you got a 'Rack' chit. If you didn't get a close enough shave, you got a 'Grooming' chit. If you screwed up during Close Order Drill, you got a 'Drill' chit. All of these were misdemeanors. The biggies were Integrity or Honor chits. Cheat on an exam and one of those chits would bounce you from the Corps. It's kinda like the Catholic Church and the concept of Mortal and Venial sin.

So after screwing up everything he touched for three or four weeks and accumulating every misdemeanor chit known to man – and all this despite heavy coaching and guidance from the former enlisted men – Sloan really made an effort one Saturday morning to outshine (literally) every other candidate.

Every Saturday morning throughout OCS, before we were granted weekend liberty at noon (and had to be back by Sunday night), we had an inspection. One week it might be uniforms, another it might be weapons. This particular Saturday morning we were all standing in front of our racks (two-tier bunks) for the 'junk on the bunk' inspection. Junk on the bunk was the precise display of our field gear. In accordance with the inspection manual, each piece of gear – pack, suspender straps, web belt, first aid pack, canteen, ammunition pouches, etc. – must be precisely laid out on top of our perfectly made bunks. Each piece not only had to be spotlessly clean, but also had to be a precise measurement from each other piece. We all hated preparing for junk-on-the-bunk, but Marine officers were supposed to be masters of detail.

Our Sergeant Instructor, Staff Sergeant Howard, was meticulously inspecting each piece of gear. He had done our side of the barracks handing out a few chits and was coming down the other side. Candidate Sloan's rack was almost directly across from Ross's and my bunks. When SSgt. Howard got to Sloan's rack, he began looking intently at each piece of gear. He snatched up Sloan's metal canteen cup and said, "Candidate Sloan, how did you get your canteen cup so shiny?"

We all looked over to see this marvel, and there in SSgt. Howard's hand was a canteen cup that looked like it was chrome plated.

Canteen cups usually looked dull and gray.

Canteen cups were supposed to look dull and gray. They were designed that way so that when you were in the field there would be no glint of reflective light off your canteen cup to give away your position and, like, kill everyone in your platoon.

Sloan proudly answered the question with, "Platoon Sergeant, the Candidate polished it with Brasso."

Upon hearing Sloan's response, Ross and I began to fight off suppressed laughter at the thought that someone would be dumb enough to clean a drinking cup with Brasso (a toxic liquid used to clean our brass belt buckles). Besides the risk of poisoning himself, Sloan would probably have diarrhea for a month. Then when SSgt. Howard drawled, "Candidate Sloan, I am giving you a Stupid Chit," Ross and I just lost it. SSgt. Howard turned to look at us and said, "And you two clowns are going to do PT (Physical Training) until I get tired."

Ross and I did endless push-ups for that.

But it appears that Sloan is not done with us yet, not by a long shot. That junk on the bunk gem was on a Saturday. The following Wednesday, week five, I thought it was all going to come to an end. SSgt. Howard was walking the platoon through a complicated sixteen- or twenty-step close order drill movement which, in an ideal world, would result in the forty-two of us candidates being able to create fourteen separate rifle pyramids in groups of three.

We started by walking through the series of steps it takes for a platoon to open ranks and then stack rifles in pyramids of three. The Marine Corps does everything by the numbers, and this drill movement was no exception. It requires detailed scrutiny from our sergeants because the movement has so many moving parts.

We had been marching in platoon formation, three columns abreast, when SSgt. Howard gave the command to halt. The next command was "Order Arms," which meant bringing the rifle from our shoulders to a position down by our sides with the butt plate of the rifle resting on the deck and the barrel of the rifle lightly held in our right hands. The next command was 'Left Face,' followed by 'Open Ranks' (first squad taking two steps forward, second squad taking one step forward, and the third squad (containing Ross and me) remaining stationary. Then there was the 'Dress-Right, Dress' command to line us up. The next command was the first of several that would result in our achieving the desired goal of being able to stack three rifles together.

At that point every three men would form a triad, with the man in the middle standing straight up with his rifle brought around to center on his body and the butt plate of the rifle on the ground. The men on either side of him were to lean over and join their rifle barrels together with the middle man's rifle, and link the rifle barrels inside the rifle slings to form a pyramid.

Simple. Got it?

At each point in the exercise, SSgt. Howard had us freeze as he walked up and down the ranks, making corrections to our positions here and there. For our team, Ross was the center man and I was on his left side, bent over to link my rifle in our triad. As we were in the third squad we could see the backs of our fellow candidates in the first two squads. SSgt. Howard stopped in front of a trio of candidates in the first squad and said to one of the bent-over candidates, "Candidate Sloan, you have a hole in your cover." (In the Marine Corps we call hats covers.)

Marine field covers, the ones worn with the camouflage uniform, are distinctive from all the other services. They resemble, remotely, baseball caps. All Marines almost daily wash their covers, wring them out, and stretch them over a cylindrical 'cover block,' paint the cover with liquid starch, anchor the cover to the block with clothes pins, then wait overnight while it dries. Once removed from the cover block the dry cover's edges can be creased and thereby given a 'salty' look unique to the Marine uniform.

At SSgt. Howard's observation, Sloan stood quickly to attention and said, "Oh, no, Platoon Sergeant, it's just that when the Candidate was drying his cover on the cover block it wasn't quite dry, so the Candidate tried to use a hot iron on the top of the cover to dry it." All of us in the next two ranks could see that the top of Sloan's cover had the seared dark brown burn imprint of the hot iron where he had left it too long. At that point, SSgt. Howard gently gripped the bill of Sloan's cover in his left hand, removed it from Sloan's head, balled his right fist, and punched through the burned spot. He then took the cover, placed it on Sloan's head, and pulled it down over his ears so that Sloan's ears were bent outboard (a la Yoda) and the iron shaped flap was sticking straight up in the air.

In his soft, upper east Tennessee drawl, SSgt. Howard said, "Candidate Sloan, I said you have a hole in your cover."

That was all Ross and I needed. We got laughing hysterically to the point where our sides were aching and we were hanging onto our rifles to keep us from completely dropping to our knees. SSgt. Howard looked at the two of us and said, "And you two idiots can work off your laughter by running around the parade deck with your rifles over your head until I get tired."

So off Ross and I went with our rifles over our heads and we ran until I thought we'd drop.

I just knew we're going to get kicked out of here.

One morning around the end of the fourth week we were all crowded into the head trying to brush our teeth and shave and then get out to formation, when we heard the strangest sounds. "Arrrrgghhhh. Arrrrggghhh! Arrggghhhhhhhh!!! Ross looked at me and said, "What the hell was that?" I said, "Beats me." Ross walked around the corner and came back a moment later and continued shaving. He said that Candidate C was in the gear locker (broom closet) barefoot, wearing civilian pants and no shirt, standing there panting and making strange noises. Ross figured C was just trying to get attention as he usually did. As Ross and I completed our toilet, we heard, "I'm going to kill myself!" Ross and I went to look and saw that C had a razor blade in his hand and was gesturing with it toward his wrist.

Just then Lieutenant Matulich, our platoon commander who also that day happened to be the Officer of the Day, came in, quickly assessed the situation and knocked C down and took the razor blade from him. He had a couple of candidates restrain C and sent for the Corpsman who quickly arrived and sedated the candidate (he should have sedated the lieutenant, too). By late that afternoon C's locker had been emptied and, following a brief hospitalization, he was medically discharged from the Marine Corps.

In week five, Candidates S and Mc were dropped from the program as unsuitable to become officers. They were sent to Parris Island for recruit training.

One anecdote about Sergeant Howard deserves note. One Friday night I was standing duty in the company headquarters as Duty NCO. My friend and fellow candidate from OCS, J.D. Richardson was acting as Assistant Duty NCO. The real Duty Staff NCO was our platoon sergeant, SSgt. Howard. As I mentioned earlier, we all had to refer to ourselves in the third person, and to our superiors as Platoon Sergeant or Sergeant Instructor or whatever.

Well, there we were on a quiet Friday night and I asked the Platoon Sergeant to tell us about his own Marine career. He began to tell us about his experiences as a junior Marine and the training he went through, and, after a pause, told us about being deployed in 1962 during the Dominican crisis. The

Marines were sent to Santo Domingo to restore order following a coup, and they were deployed around the perimeter of the main airfield. Daily at the civilian airport, the US Air Force troops had hot chow flown in, while the Marines guarding the airport perimeter were given cold 'C' rats (that's what we ate back then – C-rations which were canned meals that one would have to be very hungry to enjoy; I still carry on my car key chain the C-ration can opener – the P-38, affectionately called in the Marine Corps a "John Wayne;" it is one of the most useful tools I own). SSgt. Howard told us of the dysentery he suffered guarding that perimeter while the Air Force pukes had hot chow flown in from the United States every day.

The bottom line was that it made him more devoted to the Marine Corps.

My Colleague, J. D. Richardson, felt that a familiarity had been reached with our platoon sergeant. He said, "SSgt. Howard, …"

And he got no further. The response was, "SSgt. Howard? SSgt. Howard? My wife calls me SSgt. Howard."

Well, Immediately J.D. got flustered. I, however, began laughing, and continued to laugh until I realized SSgt. Howard was staring at me. After a moment, he gave me a nod and a slight smile, and we resumed our 'conversation.'

So much for the lighter side of OCS.

Our platoon leader, as I mentioned earlier, was 1st Lieutenant Matulich. He had been wounded in one of his knees in Vietnam and was assigned to be our platoon leader until he received his medical discharge. That was a major mistake by the Marine Corps. All officers must be able to conduct the exact same training that their troops undergo. Part of leadership, and a very important part, is that Marines see the example being set. In our case, we had an absentee leader for most of our training.

Sometimes the lieutenant seemed be a decent guy. Unfortunately we saw too little of that side of him. Mostly we saw his small and petty behavior. Because of his injury he could not lead us weekly on the Hill Trail, though he tried the first couple of weeks. Nor could he accompany us on our daily runs. Nor was he around for most of our physical training. What was most mystifying was the fact that he wasn't around for even the classroom training (it made us wonder just what it was that he did with his time).

The day-to-day leadership tasks fell to our sergeants. So be it. We had no problem with that. A medical condition is what it is. What we did have a problem with was the fact that one particularly grueling Friday afternoon we got back from the hill trail excursion tired, filthy, and just looking forward to dropping our packs, taking a shower, and having a couple of beers before we had to prepare for the Saturday inspection. Unfortunately Matulich in his clean uniform waited for us. His idea of leadership, when he had not shared our field experience, was to insult us and put us through rigorous PT while we still carried our packs.

It became evident to us, even though they masked their feelings well, that SSgt. Howard and his staff were mortified to have this idiot abusing his authority. The good platoon sergeant's solution was to protect us as much as possible from the lieutenant and give us the leadership we needed.

In spite all of that, we kept our morale pretty high and pushed on in our quest to become leaders of Marines. We knew what the ideal of a Marine leader was and that is what we all tried to live up to. Personally, I was on a quest to become an officer, and neither Matulich nor any other officer was going to deter me.

Leading one of our sister platoons was a newly minted Captain whose name was either Ingam or Ingram. I'll use the latter. He was to be my tormentor – dare I say Torquemada? -- for several weeks of OCS. I wasn't to figure out until quite some time later that he was the Officer of the Day on that Saturday when I was required to walk one mile to mainside on two bad legs. As the Officer in Charge that day, he must have gotten his ass reamed by the CO of OCS, Colonel Ing. Captain Ingram's mission from that point on was to destroy me.

The sad thing was that he never said to himself, "I screwed up. I should apologize to the Candidate and try to do better."

The first incident of his revenge came the week following the transportation snafu (probably Week 3). In both classroom and the field we were learning the art of writing out tactical orders. We initially received the appropriate amount of class time learning how to write a five-paragraph order. The Marine acronym for this order is SMEAC, which stands for Situation, Mission, Execution, Admin and Logistics, and Command and Communications. It is pretty basic stuff until you are in the field simulating a tactical situation and you have just a few minutes to prepare one.

Training us in the field, Captain Ingram was in charge of giving us a five-paragraph order from the company commander's perspective, and we were to break down the Company CO's order and derive our own Platoon SMEAC formatted order as if we were platoon leaders. One day, as soon as the allotted time was up for us to prepare our orders, he called on me to give mine, which I did, and then proceeded to rip it apart. As I recalled, his critique took a personal slant as he blatantly said that I was incompetent and would get Marines killed. Nice example of leadership: humiliate your student in front of his peers. Okay, I thought, so I did it wrong. I'll do better next time. Well, we ran through the exercise many times that day, and each time he singled me out for extremely harsh criticism.

Later on after the exercise was over, several of my fellow candidates came up to me and said that they thought my orders were as good as anyone else's and couldn't figure why Ingram was locked in on me. I just shrugged it off and went on with the program.

A week or so later, maybe week four, Ingram was the Officer in Charge of our pugil stick training. I've covered this in the Boot Camp section, so I won't repeat myself. Our platoon was doing the training alongside his platoon. As each of our turns approached, we donned the protective gear and went out to match up against someone from the other platoon who was of equal size and weight. As I mentioned earlier, the SOP for the Marine Corps during both recruit and officer candidate training required that such training be conducted by trainees of the same approximate size in order to prevent unnecessary injury. It is clearly articulated in the SOP that it is an absolute requirement that as soon as the first killing blow is delivered the match is over and the contestants must retire to the end of the line.

Both of our platoons had dutifully lined up by height and size so that we properly would be paired off. When it came my turn to step into the arena against a similarly sized opponent, Ingram pulled my scheduled adversary out and replaced him with a guy from his platoon who stood about six foot two and weighed about 225 pounds. I was five foot ten and a half and weighed 172. This guy was solid muscle and had been an All America linebacker at a Division 1 college before joining the Corps. So I waded in and gave it my best, but the mismatch was too much and this guy clobbered me on the side of my head for a killing blow. I went down with my bell completely rung. I figured that was it.

I slowly got to my feet and headed for the rear of the line, but Ingram stopped me and made us fight again.

And again.

And again.

And again.

Five times, each time with the same result.

By now my head was killing me. I was seeing double, and I had also strained a tendon in my left elbow. Later on my opponent came up to me and apologized and told me that Ingram had it in for me. Really? Ya think? Nice job telling the Captain that you didn't think the order was lawful or reasonable, you miserable fucking puke. You can take your All-America trophy and put it where the sun don't shine. That was what I wish I had said. I actually nodded and said something like 'no problem.'

I want you, gentle reader, to understand that we trainees, in both enlisted and officer programs, had been hammered with the fact that we were obliged **not** to follow such morally unlawful orders and to require that they be given to us in writing if the officer insisted.

The other candidate felt he had no choice but to follow the orders given. All-America linebacker or not, he wimped out and fell back on that old tried and true testimony used at the Nuremburg trials – "But I vas yust following orders."

That's a major problem in a hermetic training environment like OCS or Boot Camp. The candidate or recruit feels that he or she has no options even though those options have been explained and emphasized at the beginning of training.

To his credit the big SOB tried not to hit me too hard (in matches two through five). But after that first shot, it didn't matter. He didn't have to hit me hard. And as a Division I football player, he should have known when enough was enough. He himself should have thrown the penalty flag, but he showed no moral courage.

I must have had a mild concussion because it took at least three weeks for the headaches to go away. As this training event was under the supervision of another unit's leaders, none of our sergeants, and certainly not our lieutenant, were around to see what happened.

The final insult happened around the seventh week of training, coincidentally and ironically when my headache had finally abated. We had assembled as a company on the parade deck behind our barracks prepared to go on our usual Hill Trail hike one Friday afternoon. Then the word came that the company commander wanted us assembled on the front side of the barracks, a situation that we in the military always laughingly referred to as "Change One to the basic order."

Another Abraham Lincoln anecdote occurs to me here. When he was in charge of an Illinois militia unit, Lincoln had been marching them in a flanking movement, i.e., he was not marching a column of soldiers forward, but he had taken a column of soldiers, given them a left flank order and was marching them across a field. He suddenly realized that the field was fenced, and his unit was approaching a narrow-gated opening that they could not possibly pass through in their current marching configuration. He also realized that he had forgotten the command to return them to the forward marching position. Suddenly inspired, Lincoln gave the order to halt, and followed that with an order to fall-out. He then told his troops to take a ten-minute break and fall in on the other side of the fence.

We were called to attention and the word was given to fall out and reassemble on the other side of the buildings, and we were to do it silently – no talking. So all 120 of us tried to squeeze between the seven buildings comprising our company and our adjacent company neighbor to arrive at our new formation site. In the process, as we were channeled between one of the brick barracks and the metal Quonset Hut that was the company office, I got pushed into one of my fellow candidates, Richard Poore, and I said in a whisper, "Come on, Richard, let's go."

At that moment Captain Ingram, who must have been not only stalking me, but probably was the one who pushed me, grabbed me from behind and slammed me against the side of the Quonset Hut. He grabbed me by the front of my uniform and began bashing me repeatedly against the metal building while he screamed at me that I should not be talking. My helmet encased head kept banging against the metal walls. Despite the pain in my head I knew that if I took a swing at him I'd probably end up in Portsmouth Naval Prison. He finally threw me in the direction everyone was heading and I joined up with my platoon. But, Lord, I was hurting.

The next ten miles of Hill Trail just about did me in. It was an agony, and I'm not sure how I ever made it without collapsing.

After we got back from the Hill Trail, I headed to the barracks with the intention of collapsing on my rack until the Second Coming. Suddenly SSgt.

Howard called me into the platoon office. Sgt. Schuler and Sergeant Cote were also there and I could see that they were pissed. But I immediately sensed that they were not mad at me. Sgt. Schuler said, "We saw what happened out there today. If you want to press charges against Captain Ingram, we'll testify at a Court Martial on your behalf. He had no right to assault you."

I was totally floored.

I sank down into a chair without being given permission and bent over holding my head in my hands. I thought about it for a minute or two, then stood up, looked up at my support system, and said something like, "If I press charges, I'll be dropped from this platoon and put on a legal hold until the thing is resolved, and then have to start over with a new platoon to complete OCS, right?" And they all said yes. Then I said, "Look, I was enlisted, even if only for six months before I got here, and I think I understand the game. I mean I know what will happen. Whether the Court finds that I am a victim or whether they exonerate this officer, the word will get out that I am a trouble-maker who has it in for officers, and when I get to the next OCS platoon, this false reputation will make me an easy target with the new officers I will encounter."

They all looked at each other, and finally SSgt. Howard said, "You may be right, but then again everything might work out fine."

I said, "Thank you for being willing to back me up, but all I want to do is become a Marine Officer and get the hell out of here. The last thing I need is to face the wrath of other officers who think I dishonored 'the old boy network' by doing one of their own 'wrong.' The sergeants all said that if I changed my mind, they were there for me. Their validation meant more to me than the satisfaction of hanging Ingram in a Court Martial.

Sort of.

I am not one to *bear* a grudge. However, being a good Irishman, instead of bearing a grudge, I tend to *nurture* it. In all these intervening years (50+), I rarely have thought of Ingram. It is, however, in those isolated moments when I do think of him, I wish I had hunted him down and repaid him in kind.

If we had a decent platoon commander of our own, these things wouldn't have happened. One of the fundamental rules in Marine Corps leadership training is that you take care of your people. And when you are in a leadership position, you never directly screw with another leader's people. If

Matulich had not been an absentee landlord, he (theoretically) would have been there to protect me from another platoon commander.

I think -- no, I know -- that there were two things that got me through OCS. One was that during the training week, I had a wonderful support system in my fellow candidates – Steve Pfeif, Jim Forney, Wayman Bishop, Dave Bice, David Ross, Sam Gauny, and many others. The second thing was at the end of the training week, every Saturday at noon, when we were released for the remainder of the weekend, Kathy would be there to pick me up. She was now doing her student teaching in Baltimore and using my '67 Mustang to commute on weekends. Friday nights she would go home to Arlington, and Saturday at noon she would drive to Quantico to take me back to Arlington for the weekend. I loved sharing the joy of being with her family. After dinner on Saturday, Kathy and I would go into DC for some fun and relaxation. Often, before Dan Sullivan graduated from Georgetown, we met up with him and his circle of friends. Sunday morning we'd go to Mass with her family. After a noon dinner I would spend a leisurely afternoon at 724 26th Place South, Arlington, Virginia, before heading back to Quantico.

Toward the end of the OCS program, around the eighth or ninth week, we had a graded event called the Speed March Reaction Course. This little gem was a test of our physical endurance, our abilities to lead, our problem solving ability under stress, and our skill at working as a team when in a state of physical duress. In squad-sized teams (12 Candidates), with each team having a senior enlisted evaluator, the event started with a timed running of the obstacle course.

First we dropped our packs and rifles and then ran over the obstacle course. The course was a series of ten physical challenges that forced you to use every muscle in your body. There were the double high bars, the single high bar, walls to scale, logs to run across, log fences to vault, a rope swing over a mud hole, and, as a last challenge, a twenty-foot rope to climb (this last was a specialty of mine for some reason). We all got through it fine, then went back to put our gear on for the next phase – a three mile run over the Hill Trail. For me the obstacle course and first part of the run went fine, but then I failed to regulate my breathing and started to hyperventilate and get lightheaded. The evaluator, who happened to be the Company Gunny, paused the squad and, said, "You can vote to leave him or bring him with you, but the vote has to be unanimous."

Thank heavens for my buddy Steve Pfeif. He was on his way to being OCS honor man (well deserved) for our company and had all the gifts of a natural leader. Steve simply looked at the others and said, "He comes with us." Everyone nodded his head and we moved on with Steve encouraging me every step of the way. I finally got my breathing somewhat regulated. We got to the end of the run within the allotted time, arriving at the site of the final event, the problem solving part.

This particular training area looked like an unfinished storage facility. The cinderblock structure was one long central wall about eight feet high with perpendicular cinderblock walls jutting off each side about every twenty feet. It was a series of exposed cubicles each about twenty feet square. Inside each exposed cubicle were ropes, planks of various sizes, poles, and various other random accoutrements. Some of these things were usable and some were not. That was for us to figure out.

Each cubicle represented a problem to be solved – a river to be crossed, a swamp to be avoided, and other challenges. Some of the problems had solutions and some didn't. The most important thing was leadership, teamwork, and problem solving.

In our particular case, we had to rescue a "downed pilot" (a life-sized dummy wearing a flight suit, boots, and aviator helmet) who lay injured on a ledge on the other side of a "gorge." The gorge was represented by two elevated platforms about fifteen feet apart spanned only by a single steel cable. Our mission was to develop a plan, assign jobs, and bring the injured pilot to the "friendly" side of the gorge. We had fifteen minutes to do it.

While I was regaining my breath, Steve took the lead and started assigning people to jobs such as making a stretcher using our ponchos and some of the poles. Then the team tried to hang the stretcher from the cable using some rope that was left there, but the stretcher would slip to one side or the other. In a flash I saw the solution. I immediately stepped forward and took charge.

First I sent two men across the cable to give initial first aid to the "pilot" and also to lift him onto the stretcher when it was pushed over. The stretcher needed to be secured a certain way so that it would not tilt to one side or another on the cable. I knew exactly how to solve that problem and assigned two men to the task. Once the stretcher was fastened securely to the cable, I got on top of the cable in a face-down position and pushed the stretcher across to the other side. The two men there lifted the pilot onto the stretcher while I swung around and then headed back, pulling a little on the stretcher while one of the guys on the other end pushed. It worked and the team got high marks for teamwork and problem solving. Best of all for me was the fact that while I got an average grade for the obstacle course and a marginally passing grade for the run, I got top marks for leadership. Didn't think I'd make it there for a while, though.

By the eighth week, the pressure was building. One of the common characteristics of the training is that as it progressed week-by-week, many of the assignments became more difficult. Often, as cited above, we were given tasks that could not be accomplished. This was deliberate. The idea was to get the candidates accustomed to extreme mental pressure and see how they reacted to the problems they were faced with.

Late in the training cycle, maybe ninth week, on that particular Friday, we had just done about a jillion miles of Hill Trail and came back pretty filthy, not to mention exhausted. We were faced with having a major inspection on Saturday that would include a junk on the bunk inspection, a detailed scrutiny of our rifles, and a detailed inspection of the barracks. No beer tonight. It was one of those situations that in the time allotted, you could prepare perfectly for one or two events, but not all.

So the former enlisted guys put our heads together and decided to take the assembly line approach. While one half of the platoon would scrub down the barracks, the other half would take all the rifles and clean them *en masse*. This would involve doing some things that were not among the Marine Corps "approved practices" for rifle cleaning.

We got a couple of empty trashcans and one of the guys went out and bought several gallons of gasoline. A good distance out behind the barracks we disassembled the wooden stocks from the rifles, and one at a time washed down each rifle with gasoline in one of the trashcans. Gasoline was not on the approved list of cleaners, however, it was a most effective cleanser. Basically it was forbidden. But it sure did the job of lifting dirt and rust off the rifles. The next step was to fill a second trashcan with hot soapy water and scrub away the gasoline. Then it was into the showers where the soapy water was washed off with clean hot water, and the rifles were immediately wiped dry and given a thin coat of oil. A cleaning rod with a cloth patch was punched up and down inside the barrel. Another group of our Officer Candidates in the meantime, rubbed linseed oil into the wooden stocks to make them shine (linseed oil was approved).

Once the cleaning of the metal parts was completed, the rifles were reassembled and joined to their shiny wooden stocks.

The third group had scrubbed down the barracks, except for the head (bathroom) where we were doing the final cleaning of rifles. They would finish that off early in the morning before inspection.

All was going well and we were about 3/4 finished with the rifles when around 2:30 in the morning the Officer of the Day came in and busted us. The following morning all of us former enlisted, easily recognized as the ringleaders, were standing tall in front of the Company Commander. He chewed us up one side and down the other, while at the same time admitting that the assignment was designed so that it couldn't be completed. After his "period of instruction," he sent us on our way with the admonition to never do it again. All we had done was overcome and adapt. We felt pretty good about it, too.

The rest of OCS was uneventful except for our meeting with SSgt. Howard a day or two before graduation. He took us aside to the far end of the barracks and gave us a great lesson in leadership. Without condemning Lieutenant Matulich or any other of the loser officers we had seen, he gave us a talk about what it means to be a real Marine Officer. The closest he came to condemnation was when he said we should not look at the examples we had seen but look to the ideal of what Marine officers should be and live up to that.

When I got my commission I was proud to give Staff Sergeant Howard my first salute and present him with a dollar, as was tradition at that time.

April 18,1969, we received our commissions. Graduation and commissioning as a Second Lieutenant was something of a relief. I was proud of what I had accomplished, but I felt none of the deeper pride, thrill, and elation I felt at Parris Island when I was meritoriously promoted to Private First Class.

The best part of the commissioning ceremony was that both my family and Kathy and her family were there, and that her father, Captain Ed Steigelman, US Navy, swore me into the officer ranks as a Second Lieutenant of Marines "... to support and defend the Constitution of the United States against all enemies, foreign and domestic."

(Three years later I returned the favor by swearing Tony Steigelman, Kathy's brother, in as a Second Lieutenant in the Army after he had completed

Army ROTC training and graduated from college. He subsequently requested a transfer to the Navy, which took about a year, and served as a naval officer for 30 years, retiring as a Captain.

(I again had the privilege of swearing in a Steigelman when in 1982 I swore Kathy in as a Navy Reserve Lieutenant Junior Grade (Lt. j.g.). She is, as I write this, a retired Navy Captain.)

The following Monday after my commissioning I reported to The Basic School, also at Quantico.

4.

TBS: THE BASIC SCHOOL

The Basic School was, and still is, the 26-week course where newly minted Second Lieutenants learn everything they need to know about the Marine Corps, all of its support structures, and how the Administrative, Logistics, and Combat Support elements of the Corps support Infantry operations. At TBS we also learned where Second Lieutenants fit into the picture. From my current readings today, the Marine Corps has dramatically improved the training there since my time.

When I arrived at TBS, I saw all the familiar faces from OCS, plus several more. Joining our ranks were second lieutenants who had been commissioned out of the Naval Academy in 1968 and therefore were exempt from going to OCS. Now, in April of 1969, they were obligated to join into the collective consciousness of the Marine Corps. There were also a great number of lieutenants who had been commissioned upon graduation from college the prior year, having completed the Platoon Leaders Course (PLC) at Quantico – also exempt from OCS but having done its equivalent in summers between Freshman-Sophomore years and Junior-Senior years in college. A final group of officers were new lawyers commissioned as Second Lieutenants in the Judge Advocate Corps (JAG).

These officers, plus those of us who went through OCS and graduated on 18 April 1969, formed up what became 'K' (or Kilo) company, 10-69.

The First platoon was made up entirely of the JAG officers and other officers whose last names began with 'A,' 'B,' or 'C.'. The remaining four platoons were broken down alphabetically into separate units that formed our TBS company. My platoon was Fourth Platoon (letters 'M' through 'S').

Our living quarters, though vastly improved from OCS, would be laughed at by our Navy and Air Force contemporaries. They would have considered our quarters to be inadequate by their own Service standards. About thirteen years later I had orders as a Marine Corps Reserve Major to attend a two-week Strategy and Policy course at the Naval War College in Newport, RI. My roommate was a Marine Reserve Major who had served in Force Reconnaissance in Vietnam. He and I were totally satisfied with our quarters because, as Marines, we were happy to be dry and warm in a bed with a roof over our heads. We Marines aren't fussy.

Our Navy counterparts, however, deemed that such quarters requiring officers of our rank to share a room was an inadequate breach of billeting requirements. While of equal rank to my roommate and me, they successfully lobbied to have private rooms in hotels in town, all, by the way, reimbursed by their service – the taxpayer in other words.

There's your tax dollar being well-spent.

But back at TBS we were happy with clean, dry semi-private rooms. Each pair of rooms was a 'suite' that shared a common head (bathroom). My new roommate was Carl P who had gone through the PLC program while attending Brigham Young University. Carl was a devout Mormon.

I was a devout Catholic, drinker and smoker, and to Carl, I was if not the devil himself, at the very least then I was a royal pain in the ass. I had no intention of trying to change his mind.

And that was small of me. I should have been more sensitive to his attitudes and beliefs.

I have met many Mormons in my life. To their credit, they are devoted to their religion. Some, however, are a bit over-zealous. Carl made the over-zealous look like flaming liberals. He was a freakin' fanatic. And now he was stuck with me - an unregenerate smoker and drinker - who was, worse yet, a Catholic! The next 26 weeks living with him were a disaster. In the beginning we got along, but by the middle of the program we both independently had gone to our platoon leader and requested a change of roommates.

Request denied. Suck it up.

But Carl really was certifiable. During the last six weeks of our training, we were mostly in the field doing infantry tactics. After one particularly hot day of training, at sunset we were standing in ranks awaiting transportation back to our quarters. One of the guys near me said, "Hey, look at that sunset! It is really beautiful." And it really was a beautiful sunset with vibrant shades of pastels reflecting off the clouds. Just then Carl said, "That's caused by pollution." So much for an artistic moment.

Another time we were doing a live fire exercise that consisted of one squad laying down a base of fire while lying in the prone position (on one's stomach) while another squad flanked the target moving in from right to left. We were firing M-16 rifles and were instructed to keep the weapons on semi-automatic, which means that we were to fire one shot at a time. We were to shift our fires from right to left as the assault teams swept across the target area. When the objective was signaled as secured, we were to stand up and slowly march forward on-line shooting to the left of the target to drive off any remaining "enemy." It is a nerve-wracking exercise because you are using live ammunition and live people. The enlisted instructors were hyper-vigilant and we were nervous about accomplishing this exercise without killing someone.

The order came to open fire and we dutifully fired our suppressing fire on the target and shifted it to the left as our comrades were sweeping across the objective. The signal was given for us to stand up and move forward on-line to help consolidate the objective. As soon as we stood up, Carl flipped the switch of his M-16 to full automatic fire and began spraying the objective and sweeping his fire left and right down along the ranks of our squad. One of the instructors immediately called "CEASE FIRE!!!" and tackled Carl and drove him to the ground, wrestling his weapon from him.

Carl's excuse was that he just got caught up in the moment. He damned near got us all killed.

After Basic School I heard that Carl was assigned to an infantry platoon in Camp LeJeune, and that when he arrived there, he immediately ordered his platoon sergeant and platoon guide to shave their heads, as he wasn't satisfied that their haircut presentation met Corps standards. Like good Marines they first complied with his order, then requested Mast, which is the means by which a junior Marine may go over the head of his immediate superior to address a grievance. They brought an official complaint to the company commander. The CO relieved Carl of his duties and arranged to have orders cut sending him to Norfolk, Virginia where he would assist in the loading of amphibious ships (and consequently not be put in charge of any enlisted personnel).

Our Basic School Fourth Platoon leader was first Lieutenant H. He was an Engineer Officer whose units cleared roads of landmines in Vietnam.

Not a stress-free occupation. The Lieutenant was not a bad guy, but he had all the signs of an alcoholic, and that certainly affected the morale of our platoon.

We were in something of a hybrid zone as it relates to our status. Yes, we were now officers, but we were still herded around like cattle. We called ourselves "Lieutenidates."

The TBS curriculum in 1969 was structured pretty well. As a nation, we were still embroiled in the War in Vietnam. All of us anticipated that we would be sent there sooner or later. The Marine Corps is structured for combat operations. Our task at TBS was to learn how each area of the Corps supported combat operations.

Most of the first 21 weeks we spent in classrooms with the occasional field trip to see what we had studied in action. We were issued a slew of manuals that were supplemented with reading handouts each week (thank God I saved all of mine). We studied the organization and structure of the Marine Corps and then began the orientation to each area of the Corps that supported the Marine infantryman in the field.

We had classes on administration ("Typewriters in the Attack"), supply, food services ("Mess Hall in the Attack"), Disbursing ("Adding Machines in the Attack"), and so on through Motor Transport, Engineering, Tanks, Artillery, Aviation, and finally, in the last five weeks, field exercises in land navigation, patrolling, and day and night infantry attacks and defense from the squad to the company level using the supporting arms of aviation, tanks, and artillery.

It was quite a lot to absorb and we were tested weekly on what had been presented.

From the first day, our routine started at 0600 when we got up and went for a three-mile run as a platoon. The morning run soon turned out to be one of Lieutenant H.'s sick jokes. He would drink himself into a stupor most nights at the Rathskeller in the Officer's Club, and many nights guys from our platoon had to carry him back to the Bachelor Officer Quarters (BOQ) at closing time. Bright and early the next morning he would prove just what a stud he was by running us into the ground.

Our Fourth Platoon formation on the run turned into quarter-mile long, stretched out mass of runners. We looked like an army in retreat. Ross and I usually finished just barely with the main pack and there would always be five or six guys straggling in behind us. H. would use the opportunity to berate us for being wimps ("Real Marines push themselves to the limit and puke at the end").

In contrast, one of our sister platoons, Third Platoon, was led by Captain Joe Tausig. Capt. Tausig was one of the best officers I have ever seen. A Naval Academy graduate, he was the son and grandson of Navy Admirals, and the nephew of a Navy Captain (all Academy graduates) who lost his leg when his ship was blown out from under him at Leyte Gulf in the Second World War. Tausig Avenue in Norfolk, Virginia was named for his grandfather, who was awarded the Medal of Honor in the First World War for leading convoys through German submarine Wolf Packs in the North Atlantic.

This guy had an impressive lineage.

A former Recon Officer in Vietnam, Joe Tausig had all the bona-fides. Watching how he led his platoon was a learning experience. He never seemed to give a direct order. He would say, "I'll be at the obstacle course at 0530 tomorrow to coach anyone who is interested in improving his technique." And his entire platoon would be out there. It was the same with the daily three-mile run. His philosophy was that we were Marines, and as such, we could not be defeated if we act as a team. He told his platoon that he knew some were fast runners, and that some were slow. The important thing was that they all finish the run together and fresh for the next event. The fast runners would slow down and the pace would be challenging for the slow runners, but not unattainable. His platoon started together, provided encouragement and motivation to each other along the way, and always finished together.

We had competitions with all five platoons throughout Basic School, but Capt. Tausig's Third Platoon won all of them.

A particularly unnerving incident occurred during our final six weeks of training. Everything we had learned was to show us, no matter what our Military Occupational Specialty (MOS) was, that our job was to support the Infantry in some way. On this one particular evolution of training, we were going to see a tank-infantry attack supported by close-air support. The training plan called for us to execute the same maneuver later that day. Our Basic School Company was assembled (along with a company of Women Basic School officers) on aluminum bleachers overlooking a "field of battle." Demonstration troops of enlisted Marines were going to simulate a tank-supported infantry attack and call in close air support. All this was coordinated by our officer instructors.

As the "attack" was beginning, our instructors called for air support. For all of our air supported exercises, the flights were flown by Marine Officer Aviators who were attending the Amphibious Warfare School at Quantico. These pilots were mid-career officers and combat-experienced fighter pilots accustomed to flying super-sonic F-4 Phantom jets in combat in Vietnam. Unfortunately, while these officers were accustomed to the high-performance characteristics of the super-sonic Phantom, the training runs for Basic School students were flown in T-2 trainers – propeller driven World War II vintage aircraft.

On this particular morning there was a 1200 foot overcast. From our vantage point looking forward, the tank-infantry attack started moving. At the same time, over our shoulders to the right the two designated ground support aircraft came out of the clouds in a straight-forward approach and ran a simulated attack run. As they flew over the target, combat Engineers blew up planted C-4 charges to simulate a bombing run.

The first pass was very impressive. The first plane flew over the target and the explosions triggered by the Engineers blew up, then the second plane flew over with similar results. Both planes impressed us with dramatic barrel-roll-outs and we all cheered. As we looked up and to our right for the second pass at the target, the first plane re-emerged from the 1200 foot cloud cover and made its pass successfully. The second plane had veered slightly off target while coming up through the clouds after its first pass, and when it came out of the cloud cover for the second pass, it was not in line with the target. The pilot, realizing his error, did two barrel rolls to bring his plane onto the target. We were later to learn that a barrel roll in a high-performance Phantom jet costs nothing in the way of altitude, but a barrel roll in a T-2 trainer at its propeller-driven speed costs 500 feet per roll. This pilot did two rolls. He actually succeeded in the second roll and was, from our perspective, leveled off, but about 5 feet above the ground. Unfortunately at his angle, he was headed right for a 75 foot hill.

He crashed straight into the hill, and there was an explosion from which we saw the pilot ejected straight up along with debris from the plane, and then a final explosion that consumed the aircraft and the navigator. This all happened about 300 yards from where we sat in the bleachers.

We were numb and shocked. Everything ended, as far as the training was concerned. The instructors were calling for medevacs and the rest of us stood in the bleachers with a grim, metallic taste in our mouths. We had just witnessed the death of two Marine Officers – a pilot and his navigator – and we would carry that image for the rest of our lives.

The official line was, we were Marines, we experienced realistic training, and we got a taste of what we would see in combat.

Over the intervening years, I have had a periodic dream, usually about once every six months, about a plane crash. Sometimes it is a commercial airliner, and sometimes it is a propeller-driven aircraft. Think there's a connection?

One of the highlights at TBS occurred in the summer of 1969. One night, July 20, 1969, we all got back from the field following a night exercise and congregated in the lounge to watch live coverage on a 21" screen TV in the Hawkins Room (our TBS bar) of Neil Armstrong becoming the first man, and the first American, to walk on the Moon.

Toward the end of Basic School, there was one event that each class of Lieutenants had to attend: the Commanding General's Reception. Each of us, and our ladies, was invited to attend the Commanding General's reception at the Officers Club (translation: you ***WILL*** be there). For this event, being in the summer months, full Dress White uniforms were required.

We assembled at the appropriate time, in proper uniform, in the early twilight of a mid-August evening, and lined up by platoons with our ladies at our sides to process into the Officers' Club and be introduced individually to the Commanding General and his wife. Our Commanding General was (then) Lieutenant General (three stars) Raymond Davis. General Davis was a Marine Corps legend. He had commanded everything from a platoon to a Division in combat in World War II, Korea, and Vietnam. He had been awarded the Medal of Honor in Korea for his leadership at the Chosin Reservoir, a particularly grisly battle that lives in Marine Corps history.

In November of 1950, the First Marine Division had advanced all the way through North Korea to the Yalu River separating North Korea from China. The temperature dropped to 40 degrees below zero, and then eight Chinese Communist divisions surrounded the 1st Marine division and moved in for the kill. The 1st Division staged a strategic withdrawal to the sea, and in the process, annihilated the eight Chinese divisions surrounding them.

General Davis's leadership was instrumental in saving the division. In addition to the Medal of Honor, Davis had also been awarded the Navy Cross,

Silver Star (three times, if I remember correctly), and the Bronze Star twice, not to mention a couple of Purple Hearts. He had every medal for heroism in combat that the PX carried. And now we were to shake hands with him.

We were all excited to be meeting this legend and hoped that we would not screw up our introductions.

So there we were on a lovely August night, lined up in our Dress Whites, each of us sporting a single National Defense Medal (which they gave to every service member who had a pulse), proceeding slowly through the reception line. Lieutenant H. was at the back end of the line of our platoon making sure everything went smoothly. Finally Kathy and I got to the head of the line. I handed the General's aide my calling card (that said simply, Brian J. Power, 2nd Lieutenant, USMC), and gave him my fiancee's name. He, in turn, introduced us to Mrs. Davis and the General. I shook Mrs. Davis hand and we exchanged So-Nice-To-Meet-You's. I then proceeded to General Davis to shake his hand. As I was shaking his hand and he was saying, "Nice to meet you, Lieutenant," I heard Mrs. Davis say to Kathy, "Kathy Steigelman? You're not Anne Steigelman's daughter are you?" And Kathy responded, "Why yes, Ma'am, I am." And Mrs. Davis then proceeded to tell Kathy about the volunteer groups that she belonged to with Anne, and how wonderful Anne was, and on, and on, and on.

In the meantime, I am standing face to face with General Davis, and other than being fellow Marines, we had NOTHING NOTONEGODDAMNEDTHING at all in common, and all I could do, because he was taller than me, was keep staring at the Medal of Honor draped on a ribbon around his neck. I mean this guy literally carried a Marine Battalion on his back as he led Marines out of a Chinese Communist trap in forty-below temperatures at the Chosin Reservoir in Korea. Added to my angst my peripheral vision could easily see the stacks of combat awards for bravery mounted on his left breast. Like he and I had a lot to talk about. And that particular moment seemed to last an eternity, maybe even longer. He finally said something like, "It seems that my wife knows your fiance's family." And, mustering all of my wit, charm, and full conversational abilities, my dry throat choked out, "Yes, Sir."

Finally we moved on with Mrs. Davis's voice in our ears, "Be sure to tell Anne I said hello."

Like every other Lieutenant that evening, having endured and survived "the ordeal," we headed downstairs from the Main "O" Club to the basement

Rathskeller for beer. At the head of the stairs, as we were about to descend, we were intercepted by Lieutenant H, who had a panicked look on his face. Given the history of his refusing to change my roommate assignment, and as he and I weren't on the friendliest of terms, he quickly put on his best "You-Are-Now-My Newest-Best-Friend" smile, and tried to ask casually, "So what did you and the General have to talk about so long?" And I said, "Oh, not much really." And then I turned my back on him and took Kathy downstairs.

In the Rathskeller, the joint was jumping. They had a pretty good live Rock band and the place was packed. Kathy took my club card and went to get us drinks while I went to check out the band. About 10 minutes later she was back with our drinks and said, "You are not going to believe what happened at the bar." And I said, "So what happened?"

At the bar was a group of Marine Majors who were students at the Amphibious Warfare School. They apparently looked forward to these monthly receptions when all the lieutenants were there with their young wives and girlfriends. Kathy proceeded to tell me that when she went to get drinks, one of the Majors sitting at the bar tried to hit on her by using the incredibly subtle technique of grabbing her rear end. At that moment, one of my classmates, Vance Page, stepped in, grabbed the Major, dragged him off his bar stool, and told him that he was going to have to deal with a room full of my friends unless he apologized to the lady. It was no surprise that the lady got her apology and, needless to say, Vance drank free the rest of the night.

Another interesting bit of poor leadership occurred at the beginning of the last week of TBS. One of our requirements was to have our full issue of uniforms purchased before we graduated. This included dress whites, white shoes, dress blues, cufflinks, emblems, officer's sword, two sets of green "Service A" uniforms (green tunic and trousers), uniform shirts (both short- and long-sleeved), overcoat, etc. Early in our training we were advised to put ourselves on a monthly budget and purchase some uniforms each month from either the Marine PX or the Marine Shop in Quantico Town run by a former Marine.

I had gotten my issue of two Dress Green uniforms from the PX service before I completed OCS, and subsequently did all my shopping at the Marine Shop in town (their tailoring was superior to that of the PX). I was on track to have my full complement of uniforms and accessories by graduation. My mother had said she would give me the money for my sword as a graduation present and sent me a check to cover the cost. On a Monday, I deposited the check in the local Quantico bank where I had my account, ordered the sword, paid the balance on Friday, and was preparing to go in to pick it up after it had been engraved, when I was summoned to the TBS Executive Officer's office.

I got there and reported in as ordered, and the Executive Officer, a Major (and Naval Academy grad), began chewing me out for my lack of integrity and then said that they would throw me out of the Corps for writing bad checks. I said, "Wait a minute, Major. My checks are good. What's going on here?"

It turned out that the bank in Quantico notified my command that I had written a bad check for the sword. That was impossible as I not only had my pay automatically deposited in the bank every month, but also had made deposits of checks from home and had the records to prove it. And I not only told the major, I pulled out my wallet and showed him my deposit slip. He picked up the phone and called the bank, and the bank, after checking, confirmed that they had made a mistake.

The Major dismissed me without apology.

I was pissed. You don't make accusations without checking facts. Besides which, what is the matter with simply asking if I might have made a mistake in my deposits?

And what kind of a person behaves that way and doesn't see the need to apologize?

At this point in my training my morale was pretty low. I just didn't give much of a damn anymore. The only thing holding me together was the fact that on September 10, 1969, I was going to graduate from TBS and then on December 27, 1969, I was going to marry the woman of my dreams - Kathy Steigelman.

By the time I graduated near the bottom of my class I was so fed up with what I had seen of Marine officers that I began to wonder if it was worth all my effort. Nonetheless, on September 10, 1969, with Power and Steigelman families present, as well as my high school friend Dan Sullivan, I graduated and, despite class standing, was still called Lieutenant.

It's kind of like the old joke. What do you call a person who graduates at the bottom of his or her medical school class. The answer is: Doctor.

TBS became, in my limited experience, nothing more than a discordant coda to an out of tune OCS experience.

It was something I was glad to put in my rear-view mirror.

5.

NAVAL AIR-STATION PENSACOLA, FLORIDA

After graduation from TBS on September 10, 1969, my orders sent me, optimistically, to flight school at NAS (Naval Air Station) Pensacola, FL. I did not have the perfect vision to become a pilot (my vision was 20/25 thanks to a slight astigmatism in my left eye). But because I scored well on a series of standardized tests, I had qualified, at least academically, to become a Radar Intercept Operator (RIO). That meant that if I successfully completed flight school, I would be in the back seat of an F-4 Phantom jet guiding its pilot to an objective.

One of the problems I have always had in my academic life, even going back to grammar school, was that I could always score well on standardized tests, but my classroom discipline was always lacking.

So here I was now (supposedly) qualified to train and become a RIO. The RIO is the backseat driver in fighter jets (at that time the Marine Corps' choice was the F-4 Phantom, a supersonic beast of an aircraft). The RIO is the person who does all the navigational calculations, manages all the weapons systems, and handles the high-tech equipment that can keep a fighter jet from being blown out of the sky by the bad guys.

To do this job well, one must have a 'mathematical' mind, that is, a mind that can find the technical solution to the problem on short notice – like micro-second short notice. It turned out that the job of RIO was not to be for me.

Upon arrival at NAS Pensacola, we were given additional testing. A great many of us failed, and I was in some very good company. These tests had to do with physics, power plants, vectoring (whatever the hell that was), and navigation. I joined a large group of people who were now assigned to three months of what was affectionately referred to as 'stupid studies.'

Let me keep this short. I was so far in over my head that it was a blessing in mid-December to wash out of the program. Yeah, I failed. But if I had just managed to squeak by, I would have probably killed some fine pilot and myself and destroyed a multi-million dollar aircraft.

So I filled out my 'dream chit' focusing on which three areas of the Marine Corps I'd now like to go to. I dutifully put down Tanks, Artillery, and Infantry. Now was the time to get down and dirty.

The request was sent up the chain of command with appropriate endorsements recommending that I be given one of my three choices. The request probably landed on the desk of a low level government employee at Headquarters Marine Corps who had probably been denied vacation over the Christmas holidays. As a result, orders were cut, despite my failure to hack the math required to be an aviator, that assigned me to Finance School at Camp LeJeune to become a Disbursing (pay) Officer.

I thought that my prior performances at flight school clearly indicated a lack of aptitude for things mathematical. But I was proven wrong.

6.

MARRIAGE AND DEPLOYMENT

In between my Pensacola washout and my reporting to my new assignment at Camp LeJeune, NC, on December 27, 1969, at the Fort Myers Chapel in Arlington, VA, I married Anne Kathryn Steigelman. My now dear friend and former teacher from high school, Fr. Don Turlick, performed a brilliant ceremony. Afterwards so many guests, particularly senior Naval officers, who were not Catholic, said that for all the weddings they had attended, they had never been to a wedding where they had felt so much like being participants of the ceremony. I remember during the Mass Don saying to the congregation that their presence was the validation of our marriage.

Another great feature of the wedding was that my friend from OCS and TBS, Steve Pfeif, agreed to head up a team of Marine officers in Dress Blues as the lead swordsman for that arch of swords as the bride and groom left the chapel. Unfortunately the weather threw a curve into the plan. A blizzard hit the northeast starting Christmas afternoon. After an early Christmas morning wake-up in Connecticut and a hasty exchange of presents, my family and I jumped into the family car and were just able to get to Arlington, VA, after a seven hour drive, having driven most of the way in the storm (it was usually a five-hour drive).

Normally the arch of swords required at least six swordsmen. Because of the weather, only three of my TBS friends were able to make it. Fortunately Anne and Ed had friends who had a son who was a newly commissioned Second Lieutenant. He was drafted into service and the exit from the chapel under Steve Pfeif's direction was done with precision.

Kathy and I didn't really have a honeymoon, just a quick trip to Pensacola to pack up my apartment and drive north to Camp LeJeune. I was to start Disbursing School right after the New Year.

7.

B. POWER: MAN OF NUMBERS

We settled into a cheap off-base apartment at 118 Eastwood Drive in Jacksonville, NC, which we got without a lease (that should tell you something). The reason for that soon became obvious. We would only be there from early January through the beginning of March when I would get my orders reassigning me to somewhere in the Fleet Marine Force (FMF), probably, we expected, to Vietnam.

To say our apartment was a little shabby would be, as Shakespeare once said, to damn it with faint praise. But it was only for three months so we made the most of it. It had, however, been occupied by way too many Marine transients before we arrived. Wearing rubber gloves and medical masks, we scrubbed it down for days, which included removing dog hair from the refrigerator (yuccchhh!!!) and sanitizing every possible surface with Clorox bleach. Though we had moved in on a chilly January Second, we kept the windows open until the smell of bleach faded away.

On the day we were married I had seventy-five cents in cash to my name, but I had incurred no debt. I actually had that thought running through my head as we were greeting people at the wedding reception. Kathy had saved a few hundred and was also in possession of a VISA Card with a $500 credit limit. We were in love, broke, and who cared? Once in Jacksonville, we bought a color TV – using up $400 of the $500 credit limit on her VISA card. That investment allowed us to watch local programming (dominated by ads for furniture – we made furniture in the USA then, mainly in North Carolin where we were encouraged to buy bedroom "suits" i.e., "suites"; we also endured lame ads for local automobile dealers). The big prize was getting the rare opportunity to watch, in living color, TV evangelists like Ernest Angsly plying their trade. The Rev. Angsly had an amazing gift for healing sinners afflicted with a comprehensive range of maladies. The good Reverend was a one-man HMO, and all you had to do to be a part of it was to send him some of your hard-earned dollars. Unsurprisingly he could not cure himself of his baldness as it was patently obvious that he wore an ill-fitting toupee.

We also got the best college basketball in the nation as we were at mid-point in the Atlantic Coast Conferernce (ACC). Getting to see UNC, NC State, Maryland and the rest taking on the other great college programs was a pleasure.

After the apartment had been sterilized, we moved in our few possessions: a bed given to us by my parents, a couch donated by Kathy's parents, and a few other modest household goods. We started to adjust to domestic life.

We had the good fortune to reconnect with Tom Reidenbach, my friend from enlisted days at the Infantry Training Regiment. Tom was now a sergeant, married and stationed at Cherry Point, about 45 minutes north of LeJeune. We periodically, over those few months before my deployment, played host to each other for Saturday night dinners. Tom and Sandy were a lot of fun to be with. Some of my old Basic School mates, bachelors all, also dropped by for dinner and drinks. A few who were just passing through found shelter on our modest living room couch.

We made a couple of side trips while we were down there. One such trip was a long weekend drive (probably Presidents Weekend in February) up to the Outer Banks to see Cape Hatteras, Nags Head, and Kitty Hawk. It was a great weekend trip (although Kathy was not wild about the ferry boat that connected us from the mainland coast of North Carolina to the Outer Banks). We fell in love with the wild serenity of that beautiful area of the country. That wild peacefulness is now long-gone with the arrival of condos and time shares.

Another weekend trip was my pilgrimage to Parris Island. I had told Kathy all the stories, and so we decided to make a weekend of it and go down for a visit. I was hoping that I could find my old Drill Instructors to show them that one of their "Hogs" had made it to the officer ranks. Friday night on the way down, we were stopped for speeding (in our '69 VW Beetle with Connecticut plates -- are you shocked? Shocked?) by South Carolina State Trooper, Clyde P. Dingle (I kid you not). The ticket cost us $15 cash paid on the spot, no receipt offered. At that time and place in American culture, this was a classic example in the South of being pulled over for driving while Yankee. In subsequent years I developed a strong empathy for those who are pulled over everywhere for driving while Black or Hispanic.

In Beufort (b'yewfort), SC, home of Parris Island, we found a motel that met basic hygene standards and checked in for the weekend (no Trivago in those days).

Saturday morning, after a good breakfast, we entered Parris Island, this time on my terms, and arrived in the 2nd Battalion Recruit Receiving Areas around late-morning. I had elected to arrive in uniform and decided to stop by the 2nd Battalion Recruit Receiving Barracks to see if anyone there could direct me to one of the DIs.

Kathy stayed in the car while I went inside. As I stepped into the middle corridor between the two wings, immediately a Gunnery Sergeant came up to me with a panicked look on his face thinking that I was from the Inspector General's office or something. I could hear that there was a new group of recently arrived recruits being lined up and tongue-lashed at the same rows of tables in the wing to my left where I had once stood. I quickly reassured the Gunny that I was not a spy and asked him for assistance in running down either SSgt. McDowell, Sgt. Crall, or Sgt. Rivers. He told me that McDowell and Rivers were no longer at Parris Island, but that Crall was assigned to 2nd Battalion headquarters right down the street and was on duty that morning. I thanked him and left the building and said to Kathy that we should leave the car there and walk down to battalion headquarters as it was only a block away.

Along the way we saw a platoon that was just beginning to form up outside the mess hall after noon chow. As we approached the small group, they got a panicked look on their faces as they had never seen an officer this close up. They were too terrified to move, and so, encouraging Kathy to continue on, I stopped and asked them if they knew what they were supposed to do when an officer approached. One of them immediately said, "Salute, Sir!" And I told them that I had better see some sharp salutes or there would be hell to pay. They rendered their salutes, which I returned, and I then advised them to tell their Drill Instructor that they had screwed up and walked away. That was devilish of me, but it was fun, and also part of the Parris Island game.

I caught up with Kathy and we continued on to battalion headquarters. As we were walking up the front walkway to the headquarters building, SSgt. Crall came bounding down the stairs, stood to a sharp position of attention, rendered me a salute, which I promptly returned, and then shook my hand like I was a long-lost brother.

I introduced him to Kathy and we stood there chatting for a few minutes about Platoon 2017, when all of a sudden Crall turned and barked in my ear in that practiced Drill Instructor growl that I had better snap to attention when I was talking to him. My conditioned reflex from my Boot Camp experience took over and I snapped to attention. I instantly realized what had happened and broke up laughing, along with Crall and Kathy. We also could hear laughter coming from inside the battalion headquarters. It turned out that Crall had been tipped off to my coming by the Gunny in Recruit Receiving and had bet money with everyone at headquarters who would take the bet, that he could get me to snap to attention.

I was glad I could help him win that one. He invited us inside and introduced me to the battalion staff as one of his Hogs who had made good. He invited Kathy and me to join him and his wife at the Staff NCO Club that evening for dinner. He said that there were going to be a lot of familiar DI faces there that had been with adjacent platoons in the same series as 2017 when I was going through my recruit training.

We did join him at the club and had a wonderful evening. Kathy and I couldn't buy ourselves a drink, as Crall and his fellow Drill Instructors were so proud to have as their guest one of their former charges who had become an officer. I was honored and greatly humbled to receive their respect.

It was a great weekend, and one of the best social events of my Marine Corps career.

Ever.

Even with the hangover.

I finished Disbursing School which I hated, in early March, 1970. In picturing myself as a Marine, I had never imagined that I would be sitting behind a desk and operating a calculator. I was supposed to be a combat Marine living up to the traditions of those who had preceded me charging machinegun emplacements at Belau Wood, fighting off human wave assaults at Guadalcanal, slogging through black powdery sand to gain a foothold on Iwo Jima, suffering minus 40 degree temperatures while vanquishing Chinese Communist hordes at the Chosin Reservoir in Korea, and enduring with my fellow Marines the grotesquerie of the siege at Khe Sanh.

I saw myself in jungle camouflage snooping and pooping in the woods as I brought the enemies of America to their doom.

My career as a Marine Corps Officer warrior began with me as a Disbursing Officer, calculating payrolls while safely ensconced behind a desk. If there had been a course for this at The Basic School, it would have been called "Adding Machines in the Attack."

I got my orders to report to the First Marine Air Wing (Rear) in WESTPAC (Western Pacific) on April 6. First Wing was in DaNang, Vietnam, and First Wing (Rear) was in Iwakuni, Japan. We had thirty days leave, so after arranging for movers to come and pack up our things for shipment to Arlington, we headed north.

After her graduation from St. Joe's and before we were married, Kathy had gotten a government job working as a systems analyst for what was then called NAVELEX, or Naval Electronics Systems Command. She was able to return to the same job after my thirty-days leave was completed. We moved in with Anne and Ed for a few days and then quickly found ourselves an apartment in Arlington, signed a one-year lease, arranged for the movers to deliver our belongings, and moved in.

During our short time there we were able to reconnect with some of Kathy's old friends and my two friends from high school, Dan and Rick. We were able have some good times with them all. Another friend from Norwalk Community College was there and we introduced Bill Karukis to Kathy's old college roommate, Egee, and while I was overseas, they got married.

(One of the weird things in my life was that my two best friends from Norwalk Community College, Bill Karukas and Peter Fredo totally disapproved of my enthusiastic embrace of the Marine Corps. They initially blamed Kathy, daughter of a career Navy Officer, for seducing me into the military life-style. While I was overseas, my friendship with these two men really became strained. Peter, however, along with his great wife Linda, used Bill's wedding to have a long meeting with Kathy. Following that meeting, Peter sent me the nicest apology that anyone could ever hope for. I accepted it wholeheartedly and we have remained deep friends ever since. I got no such apology from Bill. It was no surprise when a few years later he and Egee divorced.

(Bill and I had connected a couple of times when I was in DC on Marine Reserve business in the early eighties and I thought that our baggage was well behind us. Then I hadn't heard from him for several years. We sent regular Christmas cards, but didn't hear anything back. Finally, when my sister Elaine passed away in 1989, I sent two very emotional letters to Peter Fredo and Bill Karukas. Peter and Linda called immediately and we had an hour long emotion filled phone call.

(Nothing from Bill. And he had once dated Elaine. Then a year later I received a scatological St. Patrick's Day card from him with just his signature. A friend's betrayal, duly noted.)

Back to 1970: our thirty days of leave were pretty well consumed with a side trip to Connecticut, getting the apartment set up and me packing for my deployment. Before I knew it, the 30 days were up and I was getting on a Pan American Airways 707 and headed for California. Kathy and I were brave with each other at Dulles Airport, but when I got on the plane and we took off, I paid $1.50 for the earphones and started listening to the airline music. The first song they played was "Leaving on a Jet Plane" by Peter, Paul, and Mary. That was when I lost it. I cried for a while, then settled in to contemplate the future. I realized later that the flight attendants had graciously granted me some space, having seen my emotional state.

When I got to Los Angeles, my first look at that part of the country, I was stunned by the women. Not a one of them was worth a second look, but the gold lame' shoes, sunglasses, and designer clothing was like having a front row seat at a costume party. I thought I had been dropped right into Halloween.

As soon as time would allow, I transferred to another airline and flew a short hop over to San Bernardino where I got transportation to Norton Air Force Base. At Norton I stayed overnight in the BOQ before getting on a chartered Pan Am flight to Okinawa. The Marine Major who was the USMC Liaison Officer at Norton was very gracious and not only patiently answered my questions, but also gave me some good "survival" tips. I also met up with several Marine First Lieutenants who would be going on the same flight. I wasn't sure how to deal with these officers who only outranked me by one rank, but they quickly put me at ease when they told me that respect between second and first lieutenants was like virtue among prostitutes.

The next afternoon we boarded the PanAm 707 stretch jet, along with about 150 other Soldiers, Airmen, Sailors and Marines, and took off. We landed in Honolulu, Hawaii, to refuel and took the opportunity to get off the plane and stretch our legs. I remember standing on the runway and being bathed with a soft tropical breeze. The mountains above the runway were shrouded in rainclouds, and the vegetation was lush and colorful. That was when I decided that this was one place I really wanted to return to. Then it was back in the air for the 9-hour flight to Okinawa.

The only highlight of the flight was that the pilot allowed those who were interested to come forward one at a time and sit in the cockpit for a bit. That was a pleasure, even if only for ten minutes, sitting there in a jump seat and looking out the front of the plane at the wide expanse of the Pacific.

We finally landed in Okinawa, which is either sub-tropical or actually tropical. It was beastly hot and humid. The Marines were all brought up to a place called Camp Hague (or Haig, I forget which spelling). Camp Hague was a dismal place. It was just a bunch of Quonset huts and warehouses, all resting on concrete slabs. The streets of this dismal camp were mud. You got from place to place at Hague by walking on wooden cargo pallets that had been laid down like sidewalks.

The "Officers Quarters" was nothing more than a Quonset hut on a concrete slab with several double decker bunk beds. Some Okinawan women would come in every day to sweep, make our bunks, and do a general cleaning for only a couple of bucks per man. Laundry service was only a buck or two more. Later in my Reserve career I learned that no self-respecting Naval or Air Force officer (I never knew about the Army) would ever have accepted such dismal quarters as being adequate. As a Marine, I was content to be in a dry place.

The "Officers Club" was a laugh. It, not surprisingly, had the same architectural Quonset Hut purity of all the other buildings. The bar was a couple of planks suspended over four upright 155 mm artillery shell casings. There were a few nickel, dime, and quarter slot machines over against a wall that I learned were staples in O Clubs outside of the United States at that time. There were a few tables where you could sit and drink pitchers of beer and eat some surprisingly good burgers. There was a slow-moving ceiling fan that did little to alleviate the heat and humidity of this tropical paradise. The only thing missing was Sidney Greenstreet in a white suit and fez sitting at a table in the corner (see the movie Casablanca to get the reference).

I reported to the headquarters building at Camp Hague and discovered that because I was the junior Marine officer on the entire island of Okinawa, I would be held for an extra two weeks to inspect troops who were coming out of Vietnam. The purpose of the inspections was to uncover contraband.

The job supervising the troop inspections was bizarre. I met with the Gunnery Sergeant in charge of this process at a warehouse where he instructed me in my duties. I was principally there to represent officer authority -- that and absolutely only that. I was to stand there displaying my Second Lieutenant's brown bar insignia (and I also had an armband declaring some kind of authority) while the Gunny and his staff handled the inspections.

This was the process: periodic flights would come in from Vietnam with Marines aboard who were headed back to the World, the Land of the Big PX, the good old U.S. of A. Assembling in this big warehouse they would form two or three long ranks. They were then ordered to dump their seabags out on the deck. The Gunny and his people, as well as some Military Policemen and their dogs, would rummage through the gear looking for contraband. Contraband was described as knives, grenades, shotguns, rifles, pistols, and various other accoutrements of war, as well as, of course, drugs.

My first experience with this came that first evening after having been assigned the job. I was awakened shortly after midnight, got into my uniform, and by 0200 was in a warehouse full of echos. Then this flight of combat Marines filed in and formed up a few rough ranks.

These guys were something to behold. Mostly younger than me, they were a lot older. They had a hard look about them. They were filthy and still in jungle combat camouflage with mud caked boots. They treated this exercise as just one more aggravation before they could become human beings again.

I just stood in the background while these guys dumped their gear out. They were fresh (or, rather, not so fresh) out of the jungle, still covered in mud and looking like they would just as soon put a bullet in the boot lieutenant's brain as anything. Remember that this was April 1970 and these people were combat veterans of some serious battles.

And you would not believe what these guys were trying to bring home: grenades, AK47 assault rifles, Light Assault Anti-Tank weapons (LAAW), shotguns, and God knows what else.

In my brief stay in Camp Hague, we never discovered any drugs. But all of the other stuff was just confiscated and locked up. No charges were brought against anyone. The troops were told just to get on the plane the next day and go home. Fortunately after we were done with them they were able to get hot showers and clean uniforms before heading back to the Promised Land.

While at Camp Hague I managed to become friendly with four Marine aviators. The only one whose name I still remember was Capt. Les Amick. He and I had a couple of adventures together. Of the others, two were F-4 pilots and one was another RIO. Within two weeks of arriving in Iwakuni, that RIO was killed along with his pilot when their Phantom had a malfunction and drove straight into a mountain. The Marine Corps is a dangerous place, even in training.

Aside from my occasional duty requirements, I had little to do. Mostly I sat around the "O" Club, drank coffee, and B-S'd with my fellow inmates who were also awaiting orders. During this time, one of the pilots I had befriended got hooked on the slot machines. As I indicated earlier, they were only nickel, dime and quarter machines. But he got seriously hooked. Over the course of one week while we awaited orders, he lost every dime (literally) he had in those machines. He kept dropping in coins and pulling that handle. I remember looking at him and he had a glassy look in his eyes and a sheen on his face. He was possessed. The experience made me never want to play those machines ever.

I only had a couple of these inspections to run before they cut my orders to Iwakuni. Some other dumb boot second lieutenant junior to me (was that possible?) had arrived to replace me.

As I had mentioned, I made friends with Captain Les Amick who was headed back 'in-country' (to Vietnam) for his second tour as a RIO.

I remember him telling me that on his first tour, after a particularly hairy mission over North Vietnam, his pilot, wanting to relieve some tension, was flying low over the ocean on the way back to base. He brought the Phantom down to 50 feet above the ocean and cranked up to Mach 1.2 (Mach is the speed of sound, a little over 700 miles per hour at sea level). I was an English major, but I think 1.2 was around 850 miles per hour. As RIO, Les could look around and he looked back and saw that they were creating a wake in the ocean as they flew along.

Les was a lot of fun to be around. On our second day there he said, "Let's get out of here and go over to Kadena." I asked what Kadena was, and he responded, "Everything this place is not."

So we made our way over the wooden walkways to the main gate where an Okinawan taxi cab took us to Kadena Air Force Base. What a shock that was. We got out of the cab at the main gate of Kadena and walked onto what could only be described as an upscale suburban community.

The roads were paved.

They had real concrete paved sidewalks and curbs.

The grass lining the sidewalks and surrounding the buildings was as green as anything you'd find in Ireland and manicured to look like a golf course.

Every building had shrubs, flowering plants and a range of greenery, all meticulously maintained.

No wooden pallet walkways or mud streets.

No Quonset Huts on concrete slabs.

In terms of personnel accommodations, the Air Force is everything the Marine Corps is not.

Les first took me to the Enlisted Club, which was really fantastic. Then he took me to the Staff NCO Club, which was better than any officer's club in the Marine Corps. Then we went to the Officers Club, which was lavish beyond belief – deep wall-to-wall carpeting, a beautiful dining area, and a long mahogany bar with leather cushioning all around it. There was also a closed-circuit television system for paging people.

I must have looked like I stepped into Oz because Les said, "The Air Force has a better appropriations lobby than the Marine Corps." We had a few beers at the bar while being eyed suspiciously by the Air Force patrons. I felt the discomfort and asked why everyone was looking at us like we were about to rob the place. Les said that there had been a brawl in the Kadena O Club about a year back and, as a result, all Marines had been declared *persona non grata* for the next six months, because they were the ones who initiated the brawl. I was to find out a little more than a year later that I was to be working for both a major and a captain who were among the instigators of that brawl.

Les told me the story as he had heard it, but when I got back to Camp LeJeune and had a chance to verify the story, my future bosses at Camp LeJeune, Major John English and Captain Dean Lucht, gave me the full scoop.

They were among a half-dozen Marine officers who had just come out of Vietnam after having spent the previous year (1967 – 1968) at Firebase Vandegrift up near the DMZ (De-Militarized Zone between North and South Vietnam). They were all Mustang officers with enlisted service going back to the Korean War. For their entire tours, they had been under constant artillery and mortar shelling across the DMZ from the enemy. When they left Vietnam for good in their dirty jungle uniforms, they flew to Okinawa.

Once there, they immediately went from Camp Hague to Kadena Air Force Base for a few drinks. While they were sitting at the "O" Club bar, some snotty little junior Air Force officer in a sharp, clean uniform stepped into the doorway and announced, "Ladies and gentlemen, please come to attention for our flying heroes!" And in walked the officers of a B-52 squadron to a standing ovation from all these Air Force pukes. To appreciate this fully, you need to know that the B-52 bombers that dropped jillions of tons of ordnance from 50,000 feet on the jungles of Vietnam, were based in Okinawa. These guys slept on clean sheets, got up in the morning, flew their missions, and then were back in Kadena in time for Happy Hour.

And you think you might have landed a good job.

John English and Dean Lucht and their colleagues thought that this was bullshit, because Marine Corps pilots live adjacent to combat zones and fly close air support at 100 feet or less above the deck to provide precision protective support for the ground troops. Doing so, they receive the unpleasant effects of intensive ground fire and anti-aircraft missiles. So Captains English and Lucht, along with their colleagues, stinking raw from the boonies of Vietnam, promptly decked the "flying hero" nearest them, and then the game was on. It was a full-scale brawl and the MPs had to be called and the Marines arrested.

The Air Force commanding general at Kadena, of course, was totally outraged, and sent an appropriately blistering message to the Marine commander on Okinawa, harrumphing in his memo (if that really can be done), that the Kadena Air Force Base Officer's Club was off-limits to Marines until further notice. The Marine commander, of course, told the boys not to do it again, then put them on the first plane bound for the U.S.

Apparently the ban had only recently been lifted when Les and I walked in for a beer. No wonder they looked at us suspiciously.

8.

B. POWER: FEROCIOUS DISBURSING OFFICER Iwakuni, Japan

April 1970 – April 1971

A day or so later, after a couple of more troop inspections, I received my long-awaited orders. I boarded a cargo plane bound for Iwakuni, Japan. No 707 stretch jet this time, just an old World War II, white-painted DC-3 that bore the Marine Corps designation of C-117. MARINES was stenciled on the side. As I bounced and jounced around for several hours, I noticed that this plane was so old that when you looked at the hatchway on the left side of the plane, you could see daylight around the door. That tells you a lot about the difference between Air Force and Marine Corps appropriations. Finally, this crate landed at Iwakuni.

Marine Corps Air Station (MCAS) Iwakuni is located on the southeastern tip of Honshu, the most populated of Japan's four main islands. Iwakuni was home to the First Marine Air Wing (Rear). Situated on a northern shore of Japan's Inland Sea, Iwakuni was about fifty miles south of Hiroshima and maybe 150 miles south of the Pacific coast port of Osaka.

At the time of my arrival, the First Marine Air Wing had already begun its re-deployment out of Vietnam. Iwakuni was where the First Wing was housed in peacetime. The Air Station was home to several squadrons of jet planes (F-4 Phantoms A-6 Intruders, A-4 Skyhawks) and helicopters (UH-1E Hueys as well as CH-46 and CH 53 transport helicopters), plus all the support squadrons required to feed, clothe, and house the 5,000 troops as well as maintain the aircraft.

My arrival in MCAS Iwakuni, while of little note, served a purpose for the Marine Corps. While Disbursing, as a necessary function of this 200,000 person war machine, might seem like a completely secondary or tertiary (fourthiary?) function in the scheme of things, it was one of the most essential elements that kept the Marine Corps machine fully functioning.

You see, morale is everything. In, I believe, the eighteen-twenties, the German analyst of war, Carl von Clausewitz, said that morale was to operations as three was to one. There is an old ironic adage in the Marine Corps that says, "There will be no money, mail, or meals until your morale picks up." Money leads the list.

In order for the Marine Corps, or any of the other armed services to function, there must be a healthy "teeth to tail ratio." That boiled down to the amount of support services needed to put the "little green man on the little green hill," i.e., keep an infantryman, artilleryman, tanker, engineer or motor transport man functioning in the field, or keep a direct support aviator in flying status.

During the War in Vietnam, I was told that the Marine Corps had the best "tooth to tail" ratio. Ours was 1 to 10. For every man in the field, there were nine backing him (or her) up in a variety of Military Occupational Specialties (MOS's).

That's where those unglamorous specialties came in. It was Supply, Mail Services, Disbursing, and a ton of other behind the scenes operations that made the Corps function.

Enter 2nd Lieutenant Brian Power, Ferocious Disbursing Officer (cape not included).

While this was not the job I wanted, it was the job I got, and I was determined to do it well.

I checked in at the Marine Corps Air Station Iwakuni Headquarters Squadron, had my orders stamped, and then was directed to a myriad of other places to be issued certain gear (.45 cal. Pistol, ammo magazines, web belt, etc,), register with the base postal service, and a lot of other places, have my orders stamped again and again, until I finally reported to the Disbursing Office and my first commanding officer, LtCol. Irene Ray. Col. Ray stood about five foot one and, to be polite, probably greatly exceeded the weight standards for

Marine officers. But she was gracious and welcoming. She directed one of her assistants to draft the appropriate letter requesting higher authority to grant 2nd Lt. Brian Power signature authority on pay records and other financial documents as required by the Navy Comptroller's Manual. That process would take about a week to take effect. In the meantime I was introduced to the pay section I would be in charge of as well as to the other 50 people in the office.

Our disbursing office had the same structure as the rest of the disbursing offices in the Marine Corps. There was the Pay Section, the Travel Section, and the Fiscal Section.

I was put in charge of a pay section that was responsible for providing accurate pay and allowances to about 1,500 Marines in six different units around the base. My Staff NCO assistant was Staff Sergeant Billy Bok from upstate New York, who turned out to be a really terrific help to me during my tour. Also in my section I had several pay clerks, including Lance Corporals (LCpl.) Moore and Carter, and Pfc. Zinni. On either side of my pay section were two other pay sections. On my left was the section run by Warrant Officer Joe Parisi, and on my right was Captain Downey's section. They were all great guys, who gave me encouragement and support and taught me a lot.

After the introductions to my colleagues and staff, Col. Ray told me to check into the BOQ and to meet her at 1700 (5:00) at the O Club for drinks.

I checked into the Bachelor Officers Quarters (the Q) and found that it was fairly decent, by Marine Corps standards (the Air Force would have submitted complaints to the Inspector General, but it was fine by all of us Marine officers). All the officers below the rank of major (or Lieutenant Commander if they were Navy - we had many Navy medical officers) were in four-bedroom suites with a common living room and a shared bathroom. I shared my quarters with two Marine Captains who were Air Traffic Controllers, and one Navy Lieutenant who was a doctor. I unpacked, took a shower, put on a clean uniform and headed for the O Club, which, conveniently, was practically next door. It was reassuring to know that from the Officer's Club it was but a short crawl to one's living quarters.

Col. Ray arrived at 1700 with several of the other officers in her wake. Captain Lamzyk, ran the travel pay section, and Captain Downey. Both Lamzyk and Downey had been Korean War veterans. I was also technically a Mustang officer, though only for eight or nine months, but these guys had many years of enlisted and war service.

I mentioned earlier that my post-Boot Camp training company CO was a mustang officer. These guys were incredible teachers of leadership and I owe so much to so many of them.

One of the most interesting things happened within a few days of my arrival at Iwakuni and left a lasting impression on me. Our Apollo 13 astronauts were in trouble in outer space. Shortly after their trajectory out of earth orbit and on-course for the Moon, their oxygen fuel tanks had ruptured and it was feared that we would lose them in space. The whole world held its breath while this drama was played out on television all over the globe.

In Rome, thousands of people assembled every day at St. Peter's Square in front of the Vatican offering up their prayers. Similar scenes occurred in Mecca, Delhi, Berlin, London, and of course Times Square. You could walk through the town of Iwakuni and never miss a second of the reporting as every storefront and bar had a TV in the window broadcasting the around the clock coverage. Even though it was all in Japanese, we could still get the drift of the situation. On base, Armed Forces Radio and TV were plugged into the major network coverage.

When the astronauts were brought down safely, I happened to be in a tailor shop just outside the base. The whole town emptied out into the streets and everyone was cheering like crazy and shaking the hand of every American they could find.

Americans were the only people on the planet that could do it, and we did.

It was a really special moment.

My Iwakuni Disbursing family was small, but all of these people had an impact on me. Warrant Officer (WO) Joe Parisi, as I mentioned, ran the pay section to my left in the office. Joe's friendship to me during my year over there meant the world to me, and I don't think I ever thanked him properly for his friendship and guidance.

Joe was a good, kind, and decent man. He was married to a beautiful Japanese woman whom he had met on an earlier tour in Iwakuni. She was from a good family, had been working in an administrative capacity on base (I believe in the Comptroller's office), and was not a former "bar girl." Joe told me the story of their formal courtship according to Japanese custom, and the difficulty he had in obtaining permission to marry. Many Marines, through just plain loneliness, fall in love with bar girls and want to get married. Joe's case was different. His intended was a nice girl who worked as a secretary on base. Joe's commanders not only held up his application for permission to marry a Japanese national, but also had him transferred to Vietnam.

Don't you love that? Hey, I got a guy in love with a Japanese woman, so let's send him to Vietnam to get shot at as punishment. Can you describe 'latent (or is it blatant?) institutional racism?'

Once in Vietnam Joe wrote to his senator, who happened at that time to be Senator Robert Kennedy from New York. Senator Kennedy, brother of the late President Kennedy, cut through all the red tape and slam-dunked through several layers of bureaucracy. Joe Parisi immediately was transferred back to Iwakuni, and he and his bride were able to marry. The Marine Corps should have learned that if it was going to screw with one of its troops, it should choose someone whose Senator is not so powerful. Joe must have also known someone at Headquarters Marine Corps, because it seemed that every three years or so, he and his wife would return for a three-year tour in Iwakuni.

I had a great time meeting all my other my new colleagues, and they told me it would be a terrific tour for me. I had my doubts.

I reported in the next morning for what turned out to be one hell of a year-long, shall we say, "learning" experience. The smartest thing I did was ask everyone to teach me what they knew. I didn't come in like some hotshot know-it-all, and that served me well. Master Sergeant Pemberton, the senior enlisted man in the office, headed up the Travel Section. That organization was responsible for reconciling all travel pay, and in a Marine Air Wing, that was a huge job.

Gunnery Sergeant Moreno was the Staff NCO in Charge of the Pay and Allowances department.

SSgt. Bok was my pay section NCO, and SSgt. Starin headed up Capt. Downey's pay section. SSgt. Ness headed up Joe Parisi's pay section.

All these seasoned Marines (and Bok, Starin, and Ness were younger than me but had a one year tour in Vietnam to their credit) taught me a ton of things: not just the technical stuff about being a disbursing officer, but more importantly the stuff about being a leader of troops deployed overseas.

There is a difference. An enormous difference.

What blew me away was the way that they looked after me and protected me.

Apparently, there had been no officer in charge of my section for several months before my arrival and, while the technical side of preparing payrolls continued, things had gotten a little loose in terms of Marine Corps standards for personal grooming and physical conditioning. Shortly after my arrival I took the section aside one day and stated my expectations of them. I expected appropriate attention to haircuts, grooming, uniform standards, and physical fitness. I then instituted a PT (Physical Training) schedule that we would conduct together, three times a week.

I let them know in no uncertain terms that they were Marines first and disbursers second. I told them that I knew this would be a challenge for them initially, but that soon they would become accustomed to, if not welcoming, it.

In short order they were looking good vis a vis haircuts, shoe shines, and uniform standards, but the physical fitness thing was a challenge. These guys sat behind desks all day and, as single guys 8,000 miles from home, spent their evenings drinking and eating junk food at the enlisted club when they weren't going into the town of Iwakuni (henceforward referred to as "the Ville") to release sexual tensions.

I just needed a little time with them.

Within a couple of weeks we were all doing the three mile run at a decent clip, and I knew that when the Physical Fitness Test (PFT) came up in June, they would be able to pass it. I tried to make the training runs a fun experience, spending time running alongside each of my men and getting to know them. The course we ran was actually pretty neat. MCAS Iwakuni sits right on a northern shore of the Inland Sea of Japan. The Marine Corps Air Station runway ran from the base of a mountain straight out over the Inland Sea. There was a seawall with a paved walkway that ran about one-and a-half miles from the main base out along the sea and ended right at the end of the runway.

We would run out to the end and watch flight ops for about ten minutes. The Phantom pilots, in prep for take-off, would, with both feet solidly on the brakes, crank up the twin J-2 engines until the plane was almost trying to take off with the brakes locked. Once given the green light from the air traffic controllers, they took their feet off the brake pedals and concentrated on holding the nose down as long as possible until they reached (beyond?) maximum possible take-off airspeed. At the last minute they would pull the stick backwards and this avionic gull-winged beast would leap from the runway right in front of us and soar nearly vertically over our heads until it achieved around 30,000 feet, what the pilot deemed appropriate cruising speed, and level off.

This was always a rush and I think my rear-echelon guys got some motivation from the experience. It put them a little closer to the reason they were providing support services.

There were, initially, a lot of out-of-shape Marines in my unit who ultimately came around to meet the Marine standard of physical fitness, but the guy in the worst shape was Pfc. Zinni. He would fall far behind the group and I would drop back to encourage him and assure him he could do it. In that process I got to know him. Within a month he was completing the runs with the group.

I think I had a pretty good relationship with my troops. One day we received notification from the Red Cross that Pfc. Zinni's grandmother had passed away back in Texas. Normally on overseas deployments, you are not given leave for a grandparent's funeral. But this was an unusual case. Although there was no official record of it, Zinni's grandmother had served *in loco parentis* when his mother died when he was a little boy. His father had deserted him and his mother when he was an infant.

Zinni was heartbroken and wanted to go back for the funeral. I spoke to Col. Ray about it and she agreed to give him a week's emergency leave. As a Pfc, Zinni didn't have the money to travel, so I loaned him $200. After he departed, the troops told me that I had seen the last of both Zinni and the $200. LCpl. Moore assured me ("With all due respect …") that I had just screwed myself. Zinni proved them wrong when he returned a week later. And he paid me back over the next six months. I felt pretty good about my investment.

As time went on and I settled into the routine, I really got caught up in my troops' welfare. On a couple of occasions while out drinking in the Ville with some new officer friends, I got some of my guys out of a couple of jams with the MPs -- just the usual drunk and disorderly stuff, and maybe an occasional fight. But I tried to look after them.

But Lord, I was a lonely person. I was writing to Kathy every day, and also sending periodic cassette tapes telling her of my life in Japan. She would also write to me every day and send me tapes as well. It was no situation to be in when you were 24 years old and married just four months. I couldn't even perceive the emotional and physical demands on the troops. I mean, take an eighteen-year old out of his comfort zone and throw him into a foreign culture where women of easy virtue competed for his affections ("I love you Marine, no shit"). I also think the weather had something to do with my emotions. Shortly after I arrived the rainy season began around the middle of April and continued through the middle of July. Jeepers, three months of rain? It was enough to depress a saint.

During that first few months I started getting letters from both my mother and my sister complaining to me about each other. In my family the tradition was never to deal with real issues directly, but instead to draft a third member into the fray to referee the match. I don't know whether that's an Irish tradition or something my family mastered on its own.

Not long after I arrived in Iwakuni, my mother sent a ten-page letter complaining that she could not understand why my sister Elaine (now aged 27) would want to leave her lovely home and move into an apartment. Elaine also sent me a ten-page letter complaining that Mom was being unreasonable in refusing to let her move out into her own apartment. They both wanted my advice on how to handle the situation.

This was just what I needed. Here I was 9,000 miles away from my new bride of four months and they were laying the responsibility for solving their problem on me. I wrote each of them back a lengthy letter. I may have used some strong terms. One letter told Mom that Elaine was 27 years old now and, despite the fact that my brother David was 32 and still living at home, it was time to cut the apron strings. I wrote another letter telling Elaine to learn how to be a little more diplomatic in dealing with Mom. I also told them both that I was having my own problems in being separated from my bride of only four months and that asking me to solve their problems from 9,000 miles away was a little over-the-top.

I received two lovely letters from them both indicating that Elaine was now going off on her own to live in an apartment in Stamford near where she worked. Mom's letter proudly pointed out the furnishings she was providing for Elaine's apartment.

Go figure.

Among my collateral duties (2nd Lieutenants always had collateral duties), I periodically stood duty as the Officer of the Day (OOD). This was a job that entailed taking 24 hours of one's weekend or holiday days to be the officer in charge of the whole base at the Provost Marshall's Office. Essentially you became the top cop of the base from 1600 one afternoon until 1600 the next afternoon.

Being the junior officer (2nd Lieutenant) on MCAS Iwakuni (all the other officers attached to squadrons were either First Lieutenants or Captains because of their pilot status), I got every holiday duty assignment. That's what you get when you are the junior person in the military and the paperwork for your promotion is lost somewhere else on the planet.

I was a little unsure of myself that first time I stood duty. I took over on a Saturday afternoon and it was a relatively quiet night as payday was not for another week. Sunday morning the Provost Marshall himself came in. Major Schmidt was a hard ass. He stood about five foot seven but had such an authoritative bearing that he seemed to tower over you. He took me into his office and, in a most authoritative manner told me that at 1500 there was going to be a surprise call-out of the Reaction Platoon and I was not to tell anyone on his staff about it in advance.

The Reaction Platoon was a rotating group of 42 Marines who had the additional assignment, beyond their regular duties, of being ready at a moment's notice to report in full combat regalia for any emergency. Reaction Platoon assignments usually were for a week's duration, then another group of Marines were given the opportunity to serve. They were a designated force to deal with riots, anti-war demonstrators, fire and flood and whatever else fell into the "Extraordinary Occurrences" category.

My job was to say nothing until exactly 1500, then tell the Sergeant of the Guard that a brig riot (simulated) had broken out and he was to activate the Reaction Platoon. If I screwed up, Major Schmidt informed me, it would be my ass in a sling. He also said that I was not to take his tone as an ass chewing, but to consider our conversation, instead, a "period of instruction."

God, how could one not love the Marine Corps?

Actually everything went as planned and the good major commended me for my proficiency in carrying out my orders.

That drill actually prepared me for a real event the following month. I was again standing OOD duty on a steamy June Sunday toward the end of the rainy season when, lo and behold, we actually did have a brig riot. Our brig held about 30 prisoners (I might be wrong, it could be as few as fifteen, but certainly not more than thirty), and most were in there for crimes like assault, desertion, direct disobedience of an order, unauthorized absence and the like. Ninety-nine percent of those incarcerated were miscreants fully capable of rehabilitation and their offenses didn't qualify as felonies. They, in most cases, did their time and ultimately became model citizens, at least until they got out of the Corps.

There was, however, one miserable human being in their midst. His name was LCpl. Frankie Davidson. This miserable puke was in the brig for making sexual advances on the Headquarters Squadron Sergeant Major's 10 year-old son. As you might correctly conclude, Frankie was not a very likeable soul. I had been on duty a few weeks earlier when he was brought in. Just being in the same room with him made me feel ill and want to take a shower.

Apparently his fellow inmates didn't like him any better, and on this hot, muggy Sunday during a recreation break in the brig courtyard when everyone was out of their cells, they all jumped the miserable deviate and proceeded to rearrange his features. I had read somewhere that even the most hardened killers in our prisons have an incredible loathing for child molesters. While the Iwakuni brig didn't remotely compare with the likes of the Attica, Shawshank, or Rahway prisons in the good old USA, our low-level detainees had just as deep a loathing for this sub-human creep.

The guards reacted immediately and went into the compound to break up the festivities. At that point the prisoners were in it all the way and started taking on the brig guards when they tried to break it up. As soon as the call came in, the Sergeant of the Guard relayed the message to me and I authorized him to call out the Reaction Platoon. The disturbance was quickly quelled, and the following day the lot of them was shipped off in chains to the Yokusuka Naval Prison (up north near Tokyo). Major Schmidt verbally commended me for taking immediate action.

Another interesting incident happened over the 4th of July holiday. Once more I was on duty (it seemed that I was designated to forever be the junior officer on base and get all the weekend and holiday details), and it happened, both coincidently and unfortunately, that not only had pay day just occurred, but also that a Navy ship was in port at the same time. Again, it was hot and muggy (it always seemed to be that way in Iwakuni, and we were just approaching the end of the rainy season which would come to a merciful end in mid-July). On this hot, steamy afternoon and night, I periodically rode out through the town in the MP Patrol van. The MPs were on regular foot patrol as well as in mobile patrol vans. Just their presence deterred Marines from getting too out of control.

I was riding with a Sergeant Robinson, a solidly built black Marine who turned out to be delightful company. We had been conversing about a range of things, and we were just turning around to head back to the Air Station when he said, "Uh, oh. We got a problem." There on the street was this really big Navy Chief Petty Officer just about to get into it with a Marine. Robinson immediately pulled over, jumped out and spun the Chief around, spread-eagled him against the side of our vehicle, handcuffed him, and put him in the back of the van.

We headed back to the PMO office to book him on drunk and disorderly conduct. Sgt. Robinson told me, as well as pointing out to me in the log, that earlier in the afternoon he found the same Chief in town, drunk and disorderly, brought him back to the transient Navy quarters (where any enlisted Navy personnel could find lodging), and ordered the Chief to stay in his room. He also told the Duty NCO in charge of the Navy quarters to ensure that the Chief did not leave his room.

When we got back to the office to book him, Robinson told the Chief that he would take his handcuffs off, but that the Chief had to stay in the PMO drunk tank overnight, then he would be returned to his quarters to pick up his gear before being returned to his ship. Right after Robinson said this and took his handcuffs off, the Chief said, "I don't take orders from niggers!" You could have heard a pin drop. Robinson just looked at him and said, "Come on, Chief, it's time you just slept it off." The Chief, who stood about six foot five and had to weigh at least 250 pounds, hauled off and punched Robinson right in the jaw. Before anyone of us could restrain the Chief, Robinson, at about an even six feet and a solid, muscular 210 pounds, just shook his head and proceeded to tattoo the Chief with a series of lefts and rights. The Chief's eyes rolled back in his head and he slumped to the floor. I just looked at the Desk Sergeant and said, "If anyone needs me, I'll be in my office."

A short time later Sgt. Robinson knocked on the door and requested to see me. He apologized, placed his .45 pistol and his badge on my desk, and said that he was prepared for me to bring him up on charges of brutality. I said, "What for?" He was a little stunned and said, "Because I beat the crap out of that Chief." And I said, "I neither saw nor do I know anything about that. What I saw was that a drunken Navy Chief stumbled and fell down."

Sgt. Robinson just grinned at me as he picked up his weapon and badge, and said, "Lieutenant, you're all right."

My relationship with the MPs paid off many times during my year there. There were numerous occasions when Marines from my office got into minor scrapes in town, and once it became known that they worked for me, they were let off the hook with warnings. Also, any time an MP needed temporary pay between paydays (a completely legal transaction, but one not encouraged by authority), I made it happen. The first time this happened, Joe Parisi was on duty as the pay officer. He asked me why I had authorized it, and I told him, "I owe him one." Joe shrugged, nodded, and continued with his duties. After that there was never a question, and I'm pleased to say that the MPs never over-did their requests.

* * *

When I first checked in to the BOQ in April, 1970, at MCAS Iwakuni, one of my suitemates was a Captain with a MOS (remember – Military Occupational Specialty) of Air Traffic Controller. I can't remember his name because I knew him for only a brief time, and we didn't hang out. He transferred home around the end of June of 1970, about three months after I got there. I asked him where he disappeared to every weekend. He told me that he and several friends traveled around southern Japan on weekends. They would go to the train station point to a place on the map and ask for round trip tickets there. I thought that that was a pretty neat thing to do. They told me of the fun they had just getting away from the base and interacting with the Japanese in a non-military environment.

I was intrigued.

Coincidentally, in late June of 1970, one of the sergeants in my office was rotating back home and wanted to sell his motorcycle. It was a really nice 150 cc Honda that purred like a kitten. The only thing he used it for was to ride off base into town for "nooners" with his young Japanese mistress. I bought it from him and decided that I would get to see some of Japan.

I enrolled in and successfully completed the motorcycle safety course required by Marine Corps Air Station Iwakuni. Throughout the summer, if I wasn't standing duty, I would get up early Saturday morning, get on my motorcycle, and head out to see the territory. I tried to choose a different direction every week. I did only day trips, but I got to see some beautiful places.

The day trip that stands out most clearly in my mind was the one where I found myself, after about a three-hour ride on a glorious Sunday summer day, going down a two-lane blacktop road through some beautiful countryside. As I was driving along, I saw a dirt road coming up on my left and I took it. The road meandered weaving around "mountains" (maybe 3000 feet high) that seemed to spring straight up from a valley floor that was nothing but rice paddies. After a while I had to slow down to make a sharp turn to the left. That's when I saw a set of stairs that had been carved into the hillside.

I parked the bike and started to climb. About 100 feet up the stairs came to an end in a little clearing about 50 feet in diameter surrounded by trees. Across the clearing was a small Buddhist or Shinto temple, I'm not sure which.

Strung over my head in the clearing was a series of wires, attached to which were little white pieces of paper with Japanese writing on them. This was a holy site where people came to leave prayers and find solitude and spiritual peace. I turned and looked back over the steps I had climbed. Out over the valley the view took my breath away. There were rice fields as far as I could see, many of them tiered up the sides of slopes and mountains. It may have been one of the most beautiful and peaceful places I have ever seen. And a large part of its beauty was its silence. That, I think, along with the view, is what really overwhelmed me. I went down on one knee and said an 'Our Father' figuring that in this holy place all prayers would be welcome.

Those trips were a lot of fun. Everyone I met was kind to me. We pantomimed to each other in our attempts to communicate. We smiled when we made sense to each other. And we laughed when we failed. I knew a few words of Japanese, and was learning more all the time, but I was light years away from holding a conversation. Often when I would step into a café for lunch, I found that the restaurants and cafes were almost always staffed with high school and college students. I enjoyed it when they would try their English out on me. I remember having a delightful conversation with a college student in Hiroshima one day. He stopped me as I was walking along and asked if I would mind him walking with me as he wanted to practice his English. I had no objection and we walked for about fifteen minutes before he had to go off in another direction. I can't remember what we talked about, but it was a fun experience.

Another source of peace and comfort to me, not to mention what it meant to my basic survival, was music. Soon after I arrived I discovered that the troops represented a new cadre of "stereophiles" previously unknown to mankind. These guys could endlessly and expertly debate (and often did) the merits of a SONY versus a TEAC reel to reel tape deck. Which receiver was best – Sony or Pioneer? I was fortunately placed in the midst of this expertise and soon took advantage of it.

Shortly after I had arrived in Iwakuni, I got to know SSgt. Ness, who was Joe Parisi's SNCO assistant. SSgt. Ness (who, incidentally, was born in Iceland) was stationed in Iwakuni on a three-year tour of duty accompanied by his wife Cathy and their little daughter. He and Cathy often had me over for dinner and he taught me a great deal about stereo equipment. We also had similar musical tastes. He had a fantastic collection of records and used to make reel-to-reel tapes of his favorite cuts from albums (this was before 8-track tapes, before cassette tapes, and well before CDs came into use).

One day SSgt. Ness took me shopping at the PX and I came out with about $1200 worth of stereo equipment for the equivalent of about 350 American dollars. The rate of exchange was so ridiculous then (the Japanese Yen was worth about 360 to a dollar). It was a shopper's paradise. I had purchased a fantastic state-of-the-art set-up that lasted us many years. The composite system consisted of a Pioneer amp, a SONY tape deck, Sonic speakers, and a Pioneer reverb unit and turntable.

I bought a lot of records and reel-to-reel tapes when I was there, and the PX offered a service where you could go in and select the songs you wanted and they would put them onto a tape. Many were the nights that I would sit and listen to the Beatles and many of the other popular groups of the day. The Beatles had just broken up, and one of their last records was the "Let It Be" album. I bought it as soon as it appeared in the PX. And when George Harrison came out with his first solo album, "All things Must Pass," I also bought it in Iwakuni and discovered that the Japanese edition was pressed on opaque red vinyl and that the liner notes were in Japanese. That became my favorite album.

Many were the nights that I put on the earphones, turned off the speakers, and got lost in George's "Beware of Darkness" and "Isn't It A Pity."

I think that one of the things that made George Harrison's music so appealing to me was that he incorporated what I had always loved in spiritual music. And that was gentle repetition. In the Catholic Latin liturgies, such repetitions happened in the litanies. The litanies were a repetitious series of responses by the congregation to a prayer intoned by a priest.

Harrison's songs provided that spiritual satisfaction at a time I was in great need.

I must have played those songs hundreds of times. I credit George Harrison for getting me through many lonely nights.

As the Fall rolled around I was given another additional duty. The Commandant of the Marine Corps, General Leonard F. Chapman, was coming for a visit as part of his annual tour of Far East bases. MCAS Iwakuni, therefore, had to prepare an honor guard for him to inspect when he stepped off the plane. I had recently been promoted to First Lieutenant (at long last – my promotion warrant had gotten lost in the international mail for three months, but the back-pay compensated for that). Again, as one of the junior officers on the base, I was tasked to be one of the 'Honor Guard' platoon commanders for the event. LCpl. Moore said, "With all due respect, it sounds like a fuck story to me, Lieutenant." Well, maybe he would be right, but I was willing to give it my best.

The other platoon commander was a First Lieutenant administrative officer from the Headquarters squadron. Each unit on the base was to send "two of their best" Marines to serve in the two-platoon ceremonial honor guard.

What a joke. Think about it. You're a unit commander and you have a requirement to send 'two of your best' troops for this assignment. 'Whoa! I have a unit to run,' the commander thinks, 'I'll send one or two of my *worst* Marines. Who's going to notice them in a large honor guard? I need my best to get the day-to-day job done.'

Well, multiply that officer by forty and you will understand just what was the caliber of troops we got for the honor guard.

It turned out that each unit on the base saw this as a way to get two of their losers off their hands for a month. I cannot even remotely perceive what these commanding officers were thinking. Here was the top officer in the Marine Corps, a man who had commanded several Marine Barracks in his career (Marine Barracks were all ceremonial 'spit and polish' operations); who was notorious for conducting excruciatingly detailed "white glove" inspections; and the unit commanders of the First Marine Air Wing (Rear), an organization notoriously lax at 'spit and polish,' are sending their worst troops out to be inspected by him? It was insane. It got me thinking about LCpl. Moore's comment.

This may very well have been the beginning of a 'fuck story.' The inspection was going to be a disaster, and I was going to be one of the officers in charge.

But I wasn't quite ready to fall on my sword just yet.

In order to assess what we would be presenting to the top officer in the Marine Corps, the other lieutenant and I, still in our naïve cocoons, assembled this sterling group one day and realized that they were sloppy looking, their uniforms, haircuts, and dress shoes were a mess. They had abandoned their Marine self-discipline, and other than that, were a sterling group to represent our command to the Commandant. Every one of them needed a haircut, their rifles were filthy, their uniforms were awful, and their personal presentations were a horror. And they really didn't give a damn about the Commandant or anything. We had one freaking serious leadership challenge in front of us.

We called them into formation and said, okay, let's start from the perspective that no one wants to be here. But here is where we all are, boys and girls. Now you get to select your options. We can have some fun for a month away from the drudge of daily routines while we prepare for this inspection, or we can send you back to your units where you will be miserable. Your choice.

Looking at the scope and lineage of great Marine leadership motivational exhortations, it was a pathetic appeal. Our exhortation may not have ranked up with two-time Medal of Honor awardee Sgt. Major Dan Daley's WWI challenge to his Marines at Belleau Wood ("C'mon you sons of bitches, do you want to live forever?"), but this was the Air Wing and it was an earnest first step.

But the troops gave us a 100 percent buy-in.

For whatever these guys lacked as Marines, they were not stupid. The consensus was, what the hell, we don't have anything else to do, plus we don't have to do our regular jobs working for idiots for a month either.

First thing we did was take an inventory. We had each of them bring their dress green "Alpha" uniform and shoes to us for inspection. The uniforms and dress shoes were in horrible shape. It appeared that each of these un-motivated troops just threw their uniforms into a sea bag without folding them (and in some cases without cleaning them). We (the other First Lieutenant as well as the Staff NCO's attached to our temporary unit) organized a massive, assembly-line-like dry-cleaning operation with a civilian cleaning business in the town of Iwakuni. We also marched the gang to the Post Exchange (PX) where there was a barber shop, a tailor shop, and a mass production operation shoe-shining operation under the direction of a local Japanese man.

And thus we began the process of turning Marines who had turned into slobs back into real Marines with pride. We also had them get regulation haircuts.

I should explain something here. Marines have a serious reputation for their exemplary appearance in uniform. Today the standards are higher than they have ever been. But back in 1970 the military services were dealing with some social circumstances that were daunting.

Vietnam was an unpopular war during a time of major social upheaval back home.

Racism in the military was becoming a ticking bomb that was threatening to turn long-simmering grievances into open rebellion.

Long-haired hippies were visible every day in the news.

Blacks in the military were angry about unequal treatment and wanted to grow Afros.

The backlash from white soldiers, Marines, Airmen and Naval personnel was intense. The quiet backlash from the officer corps that simply desired a return to the status quo was palpable. The Army and Navy now were tolerating longer hair and the wearing of peace symbols by their members. The Navy had actually authorized the growing of beards!

The Marines considered all of this as blasphemy and were trying to hold the line, but the three active duty Marine Air Wings were well-known in the Corps for being more lax on regulations than the ground forces. As for the anti-war Vietnam thing, we were on a base with only 5,000 Marines and sailors, and we had four underground anti-war newspapers in circulation.

That should give you a snapshot of the challenging military world of 1970 Marine Corps First Air Wing (Rear), Iwakuni, Japan.

Most of the officers I knew were clueless about any of this. Too much time in the Officers' Club drinking with other white officers. In short, to borrow a quote from the comedian Dennis Miller, the Marine Corps officer corps was '*dangerously homogenized.*'

The Marines we inherited for the Commandant's inspection were nothing more than a reflection of the out-of-tilt world we were living in, with some of us trying to hold to traditional Marine Corps standards.

Having convinced these troops of the benefits of being Marines for a few weeks while avoiding the nit-picking of their NCOs and officers, we took an assembly-line approach to making them presentable for the Commandant's inspection. We put an enormous effort into cleaning their rifles (ref: my anecdote about the OCS inspection). We had daily sessions with the troops on the proper maintaining of their continuous state of cleanliness. As often as we could we gave the troops close order drill. We were fortunate to have assigned to us a couple of sergeants who were fully capable of close order drill training. Within a few weeks these guys were really taking pride in themselves. They looked sharp, and they were executing drill movements with precision. They were having fun while not having to put up with the petty crap of their jobs.

No. That's wrong. There was no petty crap about their jobs. Their jobs were not petty or crappy. Their leaders were.

We also drilled these troops individually on the proper procedure for having the Commandant inspect their rifles. General Chapman was notorious for his attention to detail when inspecting weapons.

As the date approached, we developed a growing confidence, but remained anxious. Finally, the day before our inspection we got word that the General had just finished inspecting the Marine Barracks at Atsugi, Japan, and had torn them apart. Good God. If he was going to tear apart Marine Barracks Marines, what would he do with Air Wingers?

I had visions of a firing squad.

Well, there was nothing more we could do at that point. The day finally arrived and there we all were out on the edge of the runway. All the base top brass were there, and our troops were looking excellent. Shoes, haircuts, and uniforms were perfect. We had cautioned the troops that when the Commandant was inspecting them, he would likely be asking them questions like, 'Where are you from?' 'What is your job?' 'Do you like doing it?' 'Do you like being over here?' And so on. We instructed them to be honest, but respectful. If they didn't like being here, they could say so, but in respectful language.

Now here comes the man himself – handsome, about six-foot four, piercing blue eyes, and looking like he had just stepped out of a recruiting poster. Our Company Commander saluted and brought him to my platoon first. I saluted him and said, "First Platoon is ready for inspection, Sir!" And he said, "Lead on, Lieutenant." So I promptly did an about face and led him to my first squad. He methodically inspected the rifles and personal appearance of just about every other Marine in the platoon. As expected, he asked each of the Marines something about themselves. My Marines outshined themselves. One thing I did notice was that he made rare reference to his aides regarding rifle cleanliness. I think he was shocked that we were so squared away.

As the inspection moved on, he got his share of, "Well, Sir, I'd rather be back home" responses to his "How do you like it here" questions, but that was fine. The "troopies" were great, and the Commandant found no noteworthy discrepancies. He finished with my platoon, I saluted him, and he returned my salute saying, "Excellent job, Lieutenant." Then he went on to inspect the other platoon with the same results.

When he was done and had gone off with the Base Commander and the Wing Commanding General, we officers dismissed the troops with a "Well done, Marines." We also gave each platoon sergeant a substantial amount of money to buy them all some beers at the Enlisted Club.

Later that afternoon the officers went to the "O" Club for the Commandant's reception. While there, the Base Commander, a full colonel, came over to congratulate us. He said the Commandant was very pleased with the inspection and had ordered that the Marines in the inspection platoons all be given a 96-hour pass.

So now it was back to work again. I got back into the routine pretty quickly and life went on. However there were dark clouds forming over our office. A routine audit of the Fiscal Section (they are the ones who issue all the checks and maintain the master financial records for the office) turned up the fact that there had been some embezzlement taking place. Apparently one of the sergeants who had already rotated back to the states had cut U.S. Government checks to himself in various modest amounts, but totaling at least $50,000.

Col. Ray was not a happy camper. No Disbursing Officer-In-Charge wants to have the Marine Corps On-Site Exam Team down on his or her neck. But that's what happened next. A team flew in from Washington and spent a couple of weeks auditing all of the office records. They left finding no fault with Col. Ray's leadership or procedures, but they went back prepared to swear out warrants for the arrest of at least two of our former brethren.

So by Christmas we all had been subjected to fits of the Colonel's wrath as she dealt with the issue. But Christmas was looking pretty good for me. Kathy was coming over, thanks to Don Turlick giving us the wedding present of a round trip ticket for her. What a nice way to celebrate our first anniversary! Col. Ray told me in no uncertain terms that I was getting no time off other than two days to go up to Tokyo and pick her up at Haneda Airport, but I didn't care. Capt. Lamzyk had put me in contact with someone in Iwakuni who would rent us a house for the two weeks she'd be there, and I finalized arrangements for her visit.

I made reservations for us for one night at the New Otani Hotel in Tokyo. This was a brand new luxury hotel that had a revolving restaurant on the top floor. I flew up to meet her, and it was a joyous reunion. It was a life-saver to have her with me, even if only for a few weeks. We dined that evening in the New Otani's rotating restaurant where we could see a 360 degree display of the billions of neon lights in Tokyo. It was like Las Vegas on steroids, but more so. I was reminded of the adage I once heard: For the Japanese, anything worth doing is worth overdoing.

The next day we got a flight back to Osaka, then scurried to get from the airport to the railway station to get a train to Hiroshima. From Hiroshima we would take another train the remaining 50 miles to Iwakuni.

At the Osaka train station, Kathy got her first taste of Japanese culture. I was struggling getting the suitcases from the platform up into the rail car and was halfway down the aisle when from behind me I heard her calling frantically, "Brian! Brian!" I turned around to see a couple of older Japanese men "helping" her down the aisle by placing their hands on her buttocks and pushing her along. I smiled and bowed to them and explained, in my meager Japanese, that she was my wife. Well I got a lot of "gomenesai's" from them (that means 'I am sorry') as they bowed and backed their way out of our car.

Kathy's eyes were as big as saucers by that point, and I explained that the typical Japanese woman was not as tall as Kathy, did not have frosted blond hair like she did, nor did they possess anywhere near her northern and southern physical attributes. Further, I explained, they were trying to be both helpful and fortunate, hoping that today was their lucky day.

My lovely bride was not mollified.

When we got to Iwakuni, we settled into our new "home" for the two weeks. SSgt. Ness's wife had offered to take Kathy out shopping for tailor-made clothes to take advantage of the great exchange rate. Kathy was happy to do this while I was back at work, but it turned into another cultural shock for her.

Cathy Ness took Kathy to her favorite tailor in town where she could select the various styles she wanted and also select the type of fabric for her dresses.

Outside the tailor shop there were several old men seated on camp-stools talking with one another. Cathy and Kathy went into the shop and were greeted effusively by the woman proprietor. As they were talking, and under the guidance of SSgt. Ness's wife, my Kathy was selecting dress styles and colors and fabric. Meanwhile the old gentlemen who had been sitting outside came into the shop and set up their campstools.

My Kathy thought this was a bit unusual but didn't say anything until the dressmaker asked her to disrobe so she could measure her. Kathy looked around for a dressing room where they could have some privacy, but there wasn't one. Cathy Ness assured her that it was okay, and that the old men just wanted to look. So when Kathy was just standing there in her underwear, the old guys kept saying, "Takusan chi chi. Takusan chi chi." Kathy asked what that meant and Cathy Ness told her that takusan meant 'large' and 'chi-chi' meant breasts.

Kathy was a bit shell shocked following that excursion, so I took her to dinner at the "O" Club, like that should assuage the feelings of sexual trauma. Somehow we survived the ordeal.

I was able to take some time off for New Year's weekend, though not without an office incident I'll report shortly.

On New Year's Day, Kathy and I used the time to take a ferry boat from Iwakuni across the Inland Sea to visit the city of Matsuyama. The ferry trip was interesting in itself for introducing Kathy to more of Japanese culture. On the way over, she was feeling a bit woozy as she was inclined toward motion sickness. I said that I remembered that restrooms were down one or two decks below. She found her way down to the rest rooms only to discover that Japanese bathrooms are a bit different from western ones. There's just one rest room for both sexes. She walked in to find a porcelain toilet, however, it was seated at the floor level so one had to squat over it like a slit trench. There were side walls, separating each toilet, but no door at the front.

So there she was doing her business when a man walked in and, with his back to her, urinated into the men's common urinal which was conveniently located on the wall opposite the slit-trench toilets. At first she thought she had wandered into a men's room, but then, when more men came in to urinate, and other women came in to use the stalls, she realized that she was in a unisex bathroom. When she came topside to relate this latest adventure, she was as white as a sheet. I had to apologize to her for not telling her of the public restroom customs in Japan. In unisex bathrooms, as well as at bath spas, Japanese men would never even acknowledge the presence of a woman. No, that's not it. They would never look on a woman in either the bathroom or in a spa setting sexually. It just would not be polite. I think they were on to something.

My bride was not completely assured.

We arrived in Matsuyama and found a nice old hotel to stay in. No one spoke any English, but we pantomimed our way to a room for two nights. As it was late, we tried to order something to eat and finally worked out what I thought was going to be western style sandwiches. We each got two pieces of bread with cucumbers in between. At least they got the beer order right. We had a couple of Kirin beers and ate our sandwiches, then had a genuine Japanese bath. It was bitter cold outside, and the bath facility attached to our room had no heat. An attendant showed up to assist us and showed us the Japanese method of bathing.

The Japanese believe that one should soap up and bathe and rinse thoroughly *before* getting into the bath. A very civilized approach to cleansing oneself. This way, one was completely clean before getting into the hot water to embrace the complete bathing experience.

In the bath area near our room, an attendant directed each of us to strip and sit on stools while she bathed us in turn with a warm, soapy sponge. We were directed to soap and wash our own private parts while the attendant looked the other way. When we were done she rinsed us off by pouring hot water over us, and then directed us to the hot tub. The Japanese keep their hot tubs at probably around 120 degrees, which seems like nearly hot enough to steam a lobster. Your typical hot tub in the USA is regulated at 104 degrees.

After our soaping and rinsing, we eased ourselves into the deep steaming tubs where we could sit with our heads just above the water. After about 10 minutes of this, it was all we could take. We got out feeling like someone had removed all our bones, and then we went into our room and crawled in under nice thick quilted blankets and slept soundly through the night.

The next day, totally refreshed, we toured the city. At one point we came across a beautiful temple. We were admiring it from the outside when a Japanese college student asked us in English if we wished to go in. I said yes we would, but we did not want to offend anyone. She assured us that no one would be offended. We walked up the steps to the entrance and the student showed us where to leave our shoes. That was interesting. Here we were, 9000 miles from home, leaving our shoes to sit along with dozens of others belonging to worshippers at the temple. With a simple act of Faith, we left our shoes.

Our new guide demonstrated the proper way to bow and clap our hands twice before we entered the sacred structure. The place was simple and beautiful at the same time, and way beyond my powers to describe here. We spent several minutes inside with our guide, and finally departed. Under her instruction, we first bowed and then clapped twice upon our exit.

Next stop was a chair lift to the top of the mountain (probably two or three-thousand feet up) where a 15th century feudal castle stood. During my whole one-year tour, and during Kathy's three-week visit with me, our only experience with hostility in Japan occurred on our way up toward the castle summit. In ski-lift chairs, we passed, going down, a middle aged Japanese man who glared at us and pounded his right fist into the palm of his other hand as he went by. Must not have been happy with the outcome of World War II.

The castle was spectacular. The outer walls must have been 10 feet thick and made up of huge stone blocks, estimated to weigh one ton each, that fitted against one another so tightly that you couldn't even slide a piece of paper between them. Inside the castle grounds were beautifully laid out Japanese gardens, courtyards, and a winding twisting path that led ultimately to the dojo, that is, the center and highest building within the walls where some long ago daimyo (war lord) held his sway over his people and was shielded from his enemies. The structure was both an artistic and engineering marvel.

The most important souvenir of our trip to Matsuyama was a small ornament we purchased in a shop in the city. We put it on our tree every Christmas. The Japanese certainly did not mean the ornament to be a Christmas decoration, but rather a celebratory ornament of either the Shinto or Buddhist faith. Who cares? It is one of my favorites.

The following weekend we went by train from Iwakuni up to Hiroshima to tour the Peace Park, a memorial at ground zero of our dropping of the A-bomb in World War II. The outside of the park is beautifully landscaped with lovely contemplative Japanese gardens. Inside the Peace Park Museum is a series of photos and remnant displays, each with an anti-American message. Given the place and what it commemorated, I wasn't expecting an ode to American justice. However …

By the time we got through it all, with neither of us saying anything to the other, we both were sick of the relentless 'American atrocity' message. We looked at each other and expressed the same thought. We didn't start the war, and, horrendous as it was, this was the most efficient way of finishing it. As a side note, there were tens of thousands more casualties as the result of our firebombing of Tokyo earlier in 1945 than were experienced at Hiroshima and Nagasaki combined, but that seemed to be acceptable in comparison.

In any event, we arrived at the guest book where we could sign our names and leave a message. While wondering what to write we decided to look and see what other guests had written. There were more than enough "Mr. and Mrs. Harvey Blotz, Des Moines, Iowa, 'Makes me ashamed to be an American'" type of messages, but two messages caught our fancy. One was from one of the disbursing clerks in my office, a Navy guy, who signed it "DK1 (DK, in Navy parlance, stands for Disbursing Clerk) and Mrs. John Briney, MCAS Iwakuni, Japan: '7 December 1941 – USS Arizona, USS California (and he went on to name all the other ships sunk that day) – Payback is a bitch, isn't it?'" The other one was from some unknown Marine: "LCpl. Joe Smith, US Marine Corps: 'Hanoi, you're next.'"

Kathy and I had to get out of there fast before we totally cracked up laughing. We both agree, however, that the grounds of the Peace Park were exquisite, as only the Japanese can lay out such beautiful contemplative gardens.

Her visit came to an end too quickly, and before we knew it, at the end of the second week of January, 1971, Kathy was headed to Tokyo to fly back home. Oh, well, I would be home in April.

Now let me take you back to the New Year's Eve office debacle. At that time, the U.S. Government was on a calendar Fiscal Year. Today the Fiscal Year begins October 1 and ends the following year on September 30. In 1970, the FY began January 1 and ended December 31. Our office had to reconcile its balance on December 31. I won't go into all the gory details, but in the late morning of December 31, Col. Ray told me, the designated pay officer for the day, to close the books at 12:00. From that point on we would be paying on a January payroll until we closed the office at 3:00.

About an hour later, the Fiscal Officer, Warrant Officer Bill Byrne, asked her when we were closing off business. She told him it would be 1500 in the afternoon. The bottom line was that we were off in our reconciliation by $30,000. By 1700 we finally figured out the problem and she really let me have it in front of the troops. I was accused of all kinds of incompetence. Bill reassured me later that he got it all straightened out with her and finally convinced her that she gave each of us different times to close the books. He reassured her that there was no incompetence, just missed communications. She grudgingly accepted what he told her, but I never got an apology.

If there has been a running theme of my Marine narrative, that put one of the exclamation points on it.

A few times each year, each foreign deployed Disbursing Office needs to replenish its supply of U.S. currency. Every pay day we paid the troops in American dollars. But they were forbidden to spend American money in the Japanese economy. They could go to the various clubs on base (Enlisted, Staff NCO, or Officers) to exchange American green for Yen, which they could spend outside the gates or at the base clubs. Over time, as people rotated home, they would take certain amounts of the American green with them, and we would have to rebuild our supply. That entailed a trip with armed guards up to Yakota Air Force Base outside of Tokyo. Yakota acted as our Federal Reserve Bank.

Col. Ray decided it was my turn to make the trip. In late February of 1971, I was to take four armed guards, so I chose SSgts. Bok and Starin, LCpl. Carter, and LCpl. Moore. Moore punctuated his receiving of this honor with his usual sardonic comment, "Sounds like a bullshit story to me, Lieutenant."

Nevertheless, we boarded the Marine Air Wing workhorse C-117 transport, piloted by a lieutenant colonel, and flew to Yakota Air Force Base where we were met by an armed Air Force detachment and brought to their central disbursing office. I identified myself and presented the U.S. Treasury check for $6 million and the Air Force folks dutifully wheeled the cash out of the vault. My troops and I had to count every dollar before I would sign the receipt.

In such transfers of cash, the requirement was that each pack of bills -- ones, fives, tens, and twenties – with each containing 100 bills of each denomination, had to be counted by three different people. Once the value of each pack was determined to be accurate, the counter had to initial the wrapper. The way we counted the money was interesting. You held the pack of 100 wrapped bills in your left hand, middle three fingers behind the pack with thumb and pinky in front, left thumb ready to sweep the bills downward, and left pinkie anchored to the top of the pack of bills. As you counted, you wetted both your left thumb and right thumb on a sponge, and flipped the bills back toward yourself using your thumb. You re-wetted your right thumb as often as needed.

It took us several hours to count all these denominations, and when we were done the money was placed in several -- maybe eight or ten -- leather strong boxes about 2 feet cubed. We loaded the cash into a van that brought us out to the runway. After unloading the van, we sat on the boxes for several hours while our WWII vintage plane was receiving some desperately needed maintenance.

The troops' conversation naturally turned to how they could steal all this cash and how they would spend it. I laughed and told them that if they could haul this treasure off the base, cleverly disguise themselves as Japanese Nationals, and blend in with the population to make a safe getaway, they were welcome to try. Finally the plane was ready and we returned to Iwakuni where we had to re-count every damned bill and verify the cash before securing it in Col. Ray's vault. That was one long freaking day

I think it was in March when the biggest office crisis came up. It was a payday, and as usual, at around 0600 each squadron sent a designated pay officer and a guard to count the cash (an exact amount of twenties, tens, fives, and ones that made up each payroll), and sign for their payroll. The officers were to go back to their squadrons, pay their troops, and return the signed payrolls that afternoon.

I happened to be the designated pay officer for the day and all the squadron payrolls were in my safe. I arrived early with SSgt. Bok and set everything up for the individual squadron pay officers to account for and sign for their unit's pay.

On this particular payday, late in the afternoon the various pay officers were bringing in their payrolls, each dutifully signed by every individual Marine. One of the squadron officers who had come back with his guard said that he thought there might be a problem. Apparently after the officer had paid a Sergeant Ashford, the pay officer's pay guard said to the lieutenant that he didn't think that the guy's name was Ashford as he knew the guy from Boot Camp, and his name was something else. The particular squadron was not one of the ones that my section serviced, so I called Joe Parisi over as the squadron was one of his. He heard the story out, then called to one of his clerks, "Sgt. Hawkins, bring me Ashford's pay record from XYZ squadron." Sgt. Hawkins then said, "I'll do it in a moment, Sir, but first I have to drop something off to the Travel Section."

So we waited a bit for Hawkins to return. After an uncomfortable period of time, Joe went over to Travel to see what was keeping Hawkins. The travel clerks said that Hawkins left about ten minutes earlier and seemed to be in a hurry. With that, our whole world went crazy. Col. Ray was notified, the MPs were called, and the investigation began. LCpl. Moore said, "Sounds like a fuck story to me, Lieutenant, but then, again, it's a small price to pay to serve with the world's finest."

It turned out that Hawkins was in collusion with a squadron administrative clerk, along with their third-party buddy who was posing as the non-existent Ashford. The way the system worked back then, before everything was computerized, each disbursing officer kept a stack of blank pay records in his own safe. Whenever an old pay record was filled up with entries, or damaged (coffee spills qualified), or whenever we had to open a brand-new pay record because some Marine's pay record had been lost in the military mail system when he transferred from one command to another, we opened up a temporary pay record. These were secure documents and had to be controlled. That is why each of us kept them in our safes, and we were never to leave an open safe. We were to lock it up whenever we got more than a few feet away from it.

The same went for military identification cards. They were locked in an Admin Officer's safe at the unit's headquarters and protected. At some point, one particular Admin guy, and friend of Sgt. Hawkins, found an opportunity. His officer had left his safe open and unattended for a moment or two and the clerk stole a blank ID card.

At some point, Joe Parisi had briefly left an open safe unattended just long enough for Sgt. Hawkins, who had been patiently waiting for that very opportunity, to steal one blank pay record.

The admin guy created a personnel record for this fictitious Marine, soon to be named Ashford, complete with a record of promotions and transfers to create an aura of legitimacy. Hawkins created the pay record on which the mysterious 'Ashford' would be paid some extraordinary sums. Hawkins figured a way to slip the bogus pay record into at least two or three of the other squadron pay records in his section. This way the fictitious Sgt. Ashford would get paid on anywhere from two to four different payrolls each payday.

The unsuspecting pay clerks for each of these units would have no way of knowing that 'Ashford' was about as real as Jimmy Stuart's friend "Harvey." All they would do on the eve of payday was look at the line on the pay record that indicated pay due and add it in to their payroll. A diligent Sgt. Hawkins watched for the moment when one clerk finished his payroll and went off for a cup of coffee. Then he pulled 'Ashford's' record and slipped it into the files of a different clerk's squadron.

In addition to the regular pay-day scam, Sgt. Ashford had been paid a few $10,000 re-enlistment bonuses, each one with him represented on a different squadron's payroll. Sgt. Hawkins and his compatriot knew that they couldn't pay Ashford on the squadron payroll where the actual Marine who was pretending to be Ashford was. The chance of discovery was too great. The pay officer or his pay guard would know that this guy didn't belong in that particular squadron. No problem. Sgt. Hawkins, as one of the more experienced pay clerks, handled the payrolls of several squadrons, so he initially put Ashford on the payrolls of three of them.

It worked for a time, but, as in most cases of fraud, their plan collapse because of greed.

At the peak, when it was all discovered, "Ashford" was being paid on at least three different payrolls every two weeks, not to mention the re-enlistment bonuses. And this had been going on for months.

So on this particular payday It really hit the fan, I want to tell you. It's bad enough that Col. Ray, being of a certain age, was having some obvious difficulties with the change of life (temperature change; mood change). She routinely had periodic rages, even when she was not being investigated. Now she was facing her second Disbursing On-Site Exam Team audit in six months. That is a career killer for most disbursing officers. It really was an ugly atmosphere in that office until I departed in April of 1971.

Soon I managed, despite my continuing attempts to practice good leadership, to get myself into a pile of trouble with Col. Ray.

In March, 1971, a month before I was due to return to the States, I was given the assignment to deploy to Okinawa to pay a squadron that was on detached duty there at the Naha Naval Air Station. This was a routine occurrence in the office as squadrons were often deployed. Usually Capt. Lamzyk took the assignments to the Philipines so he could ride in the back seat of a Phantom jet. But this deployment was only to Okinawa and the mode of transportation would be the old C-117, that same reliable old crate that got me to from Okinawa to Iwakuni in the first place.

I was required to have a pay guard with me so SSgt. Bok was given orders to accompany me on my mission. We got temporary duty orders cut and took the plane south with our personal gear, two M1911A1 Colt .45 semi-automatic pistols, two magazines each loaded with six .45 caliber bullets, and a locked ammo box with about $30,000 in cash. We boarded our transport expecting nothing more than a long, routine flight.

On our flight south, the wind had picked up and soon we were bouncing around in some serious headwinds.

As we were lining up to land at naval Air Station Naha, Okinawa, the cross winds were pretty severe – gusts of 25 knots or more coming from right to left. We were sitting in sling seats on either side of the plane. I was on the left side and I could see the runway coming right at me as the pilot had turned the nose into the wind. We were approaching at about what seemed like a 45 degree angle. As we got close to the runway, one wheel touched down and then the pilot surged the engines jerking the plane to the left and the plane leaped back into the sky.

Around we went again. Same result. Over the intercom, the pilot told us that he thought he could make it the third time. Well, he couldn't. The winds were way too severe. Our last 'crash-and-dash' event left us feeling like we were doomed to splatter over an Okinawan runway (this was only the first of at least three Military aviation landing experiences which I thought would be my last moment on earth).

So we got re-routed to Kadena Air Force Base which had a larger number of air strips oriented so that you could land directly into the wind.

After a safe landing, SSgt. Bok and I, along with my locked ammo box with the $30,000 payroll in it, got a civilian cab to Naha. I can't tell you how many Naval Comptroller Regulations I broke with that decision. But I had no military transportation available, so I did what I did.

The deployed squadron commander said that payday would be the next morning at 0800 in a hangar at Naha. We tried to check in to the Officers Quarters and Staff NCO quarters, but because of heavy transient military action, there was no room for us at Naha. SSgt. Bok and I got a cab and went back to Kadena to discover that the SNCO quarters were booked solid. The bottom line was that SSgt. Bok was facing the grim possibility of having to find a civilian hotel.

At the Kadena Bachelor Officers' Quarters, I managed to secure a room with twin beds for myself and asked if there were other officer accommodations available for some of my colleagues who would be coming in later. Fortunately the quarters had a number of openings. I assured the desk clerk that a fellow Marine officer would be joining me shortly.

Believe me, Okinawa in 1970 was no place to tell a young Marine Staff Sergeant that he had to take his chances in the local village.

Back then there was no Trivago. There were no Holiday Inns or anything else like that overseas. If I didn't do something, SSgt. Bok would have to go down to "BC" Street, a sleazy nightmare of a downtown area designed to extract Marines, Sailors and Army soldiers from their money. So I made a decision. I checked into my room at the Kadena BOQ. I had previously arranged that SSgt. Bok would meet me not far from the BOQ. I changed into T-shirt and khakis and slipped out of the BOQ. I had taken my dress shirt with the lieutenant's bars on it and gave it to SSgt. Bok and told him to go check in and ask to room with me. I mean we were lugging around an ammo box with $30,000 in cash payroll along with two loaded .45 caliber pistols, and, well, it just was getting late and I needed to solve the problem.

I made a decision. Problem solved.

The next day, after a good night's sleep, we went to Naval Air Station Naha (this time with transport provided by the Air Force). The squadron Admin people had a table set up in one of the hangars for me so I could pay the troops. SSgt. Bok stood at the end of the table at the "parade rest" position – hands clasped at the base of his back, feet spread shoulder width. He and I were armed with our service Colt .45 caliber M1911.A1 semi-automatic pistols. We each had a magazine in our pistols, but had not chambered a round.

I opened the ammo box and took out the payroll. One of the squadron gunnery sergeants, thinking that he might have a future in stand-up comedy, reached over toward the stack of cash pretending to take it. In the blink of an eye, and as fast as Shane (go look it up), SSgt. Bok drew his .45, pulled back the slide to chamber a round, and now had his cocked pistol aimed right at the Gunny's temple. "I don't think you should do that, Gunny," he said.

I thought the Gunny was going to piss himself..

"I'm sorry, Staff Sergeant, I was only kidding around!!" I gently intervened and told Bok to ease up. Bok slowly raised the pistol toward the ceiling, touched the magazine ejection button, ejected the magazine, slowly cleared the chambered round, placed the round back in the magazine, and then re-inserted the magazine back into the pistol.

After that things went smoothly and we got everybody paid. We went back to Kadena and boarded the bird to fly back to Iwakuni.

Unfortunately, upon our return to Iwakuni, Bok told some of his buddies that I had gotten him into officers' quarters. Bottom line, the word soon got back to Col. Ray. She was pissed and hauled me in for an ass chewing. But I told her I had no choice; if I hadn't done what I'd done, SSgt. Bok would have to find lodging out in the civilian economy while carrying a .45 caliber weapon.

Let me tell you, Okinawa was, at that time and hopefully not now, a hell hole of massage parlors, brothels, and bars. There were local Japanese gang punks that would steal the eyes out of your head if you weren't looking. I have thought about it long and hard many times over the years and I do not regret my decision. It was either send him out to BC Street or get him into the BOQ. For me it was a no brainer. I was just taking care of one of my men and looking after the payroll. Col. Ray remained unconvinced and pissed off as well.

If Lt. Col. Ray had cared one fucking whit, she would have ensured that before she deployed an officer and a pay guard with two armed pistols and $30,000 in cash to a remote location, both of her people should have been guaranteed adequate temporary housing.

Things came to a head between LtCol. Ray and me during my last week on base. As I had filled out the paperwork when I arrived to give me signature authority on pay documents, so now, a week before my departure in early April, 1971, I had to sign papers relinquishing that authority. I could still audit pay records and initial them, but I could not sign them authorizing payment, nor did I have custody of disbursing funds and could not pay anyone.

No safe full of money.

No pay records.

No troops to supervise.

I easily could have done without the first two. But the third, leading troops, meant more than everything to me.

The last week in any command is filled with administrative trivia. You have to go around base and check out of everywhere – to Supply to return gear you never checked out in the first place; to the Armory (a legitimate stop to return the .45 cal. Pistol, cartridge belt, magazines and ammo); to Admin (to get orders cut and finalized); to Housing (to arrange packing and transport of worldly goods); etc. While doing this, in my spare time I would go into the office and ask the various officers if I could help them with auditing, a function I was still allowed even though my signatory authority was suspended. Usually the answer was no, but thanks anyway. Occasionally I was able to lend a hand here and there.

Then came the day before I was to depart and Col. Ray called me into her office. She had written my "Transfer of Command" fitness report and wanted to go over it with me. Well, disregarding a year in which my pay section had achieved the lowest payday error rate in our office; a year in which my pay section was the only unit to have achieved a 100 percent passing grade on the PFT (Physical Fitness Test); a year in which my leadership of the honor guard for the Commandant's inspection earned congratulations; Col. Ray wrote the whole thing in a fit of anger over her bizarre perception that I had not come into the office to help out during my final week. I felt like I was at the Mad Hatter's Tea Party. It was the kind of fitness report that, in accordance with Marine Corps regulations, I had to sign acknowledging receipt and also that I could attach a rebuttal statement to if I wished.

I was floored. My pay section had routinely held the lowest error rate in the office, so my performance of duty should not be in question. My pay section was the only unit in our organization to achieve 100% on the Physical Fitness Test. Also, as I stated above, I did come into the office to help out during the last week. She was really off the rails. I told her all this and she said that I could opt for attaching my own statement to the report. I grabbed the fitness report, went into the office, got hold of an available typewriter, and typed up and attached a stinging and well-documented rebuttal to her report.

The most bizarre kicker to this episode was her final words to me: "Now I want you to come over to the Club at 1700 for drinks. All the officers and I want to treat you on your last night here."

(Color me buzzing and spinning.)

I did go to the Club because I am not inclined to turn down a free drink. Plus, it was a good way for me to say a final thank you to a great bunch of guys who really taught me so many things that year. As far as Lt. Col. Irene Ray was concerned, well, she could pick up the tab and then go fuck herself.

My best goodbye came, of course, from the troops. Lance Corporal Moore drafted up a humorous fake award and gave it to me with all the other guys present. I was very touched. It was a Mr. Roberts moment (check your classic movies for the reference). I will always remember Moore's words as he said, "It's just one good fucking deal after another, isn't it Lieutenant?"

Late in the first week of April 1971, I boarded a Flying Tigers Airline (read CIA) flight in Iwakuni to fly back to the "World." (One of the weird aspects of this flight was that when I got on board, I realized that the flight had originated in Saigon, Viet Nam, where the US Army held sway; every single enlisted Army soldier on board was wearing the Bronze Star with Combat "V" for valor. The Bronze Star is the fourth highest award for combat heroism. It is an individual award. First is the Medal of Honor; next is the Navy Cross (or in the Army the Distinguished Service Cross, and in the Air Force the Distinguished Flying Cross); then the Silver Star; and finally the Bronze Star for bravery. That award should have been rarely given. And we had a plane full of Bronze Star awardees????? As we say in 2021, WTF?

It was bullshit. Yeah, a lot of these soldiers had served in combat, but just because they got dirty and shot at was no reason to give them the nation's fourth highest award for combat bravery. The Army had cheapened this great award in a tacit admission that its continued presence in Vietnam was a waste of human capital. It was just like we later cheapened youth sports: everyone gets an award. You don't have to win, but we will enthusiastically give you the same award we give to champions. Bullshit. I had a whole airplane-full of Bronze Star awardees. It was shameful. The Army was handing these out like door prizes. This was at the same time that the Marine Corps had been accused of being stingy with its bravery awards.

Taking off from Iwakuni, otherwise, was a surge of pleasure.

Nine hours later we landed at McCord Air Force Base in Seattle. From there it was a cab ride to the civilian airport, then a cross-country flight to DC. I had 30 days leave coming to me before I was to report to Camp LeJeune.

9.

CAMP LEJEUNE (AGAIN) AND ANOTHER DEPLOYMENT

It was great to be back home. Kathy left her job with NAVELEX and we headed north for a few weeks of leave. We of course stopped in Stamford to see my sister Elaine (Mom, Dad, and brother Dave had relocated to Miami by this time), then it was on up to Massachusetts where we visited historical places like Lexington and Concord and walked the Freedom Trail in Boston. Don Turlick had loaned us his apartment in Boston near where he taught at Emanuel College.

One night at Don's apartment we sat up talking late with him and the conversation got around to the Marine Corps. Don, though an ardent pacifist, never judged my choice, but never, knowing me as well as he did, could figure out why I would join such a combat-tough outfit. I told him about Boot Camp and all that I had learned there about living each day for one's fellow Marines, sharing each other's burdens, and being ready to sacrifice ourselves on behalf of our fellow Marines. He was totally surprised with my depiction of Marine Corps life and said that I had described perfectly what the ideal Christian community should be. I have to give that man credit.

It was a great vacation, but time was running out and we had to head back to pack up the apartment and head for Camp LeJeune.

We arrived at LeJeune and luckily found a nice place off-base at the Azalea Gardens apartment complex. These were relatively new apartments and were also affordable and convenient to the base. I checked in at the main base disbursing office and met Major John English who was the head Disbursing Officer for the base. He assigned me to Camp Geiger to work in the satellite office there under Captain Dean Lucht. (You may recall that English and Lucht were two of the officers involved in that brawl at the Kadena officers club; the Marine Corps sometimes can be a small place).

Camp Geiger, Infantry Training Regiment for recent Parris Island graduates, was where I had done my infantry training following Boot Camp, and where I had awaited orders for OCS.

Capt. Lucht was a great guy to work for. He was very low-key and didn't let the small things bother him. The Geiger office was there to provide pay services to the officers and men of the Infantry Training Regiment – the same place I had graduated from in the fall of 1968. It turned out that one of my good friends from OCS and Basic School was now the commanding officer of November Company, my old Infantry Training company. 1st Lt. J M was delighted to see me and asked me to address his troops so that they would know that someone had once done what they were doing and was now an officer. I gladly complied.

Kathy and I had J over to our apartment for dinner many times. He was a single man and as a result of his experiences in Vietnam as an infantry officer, he, like many of his peers, was fighting some internal demons. One evening, after a few drinks, he told us about the time when one of his patrols was ambushed and one of his men was hit pretty badly. He held the boy in his arms and the young Marine looked up at him and said, "I'm not going to make it, am I, Skipper?" And J looked at the Marine and said, "No, son, you're not going to make it." And the Marine looked up at him and said, "Tell my Mom that I love her." And then he died.

America expects a lot from its newly minted young twenty-something year-old officers whose recent work experience was teaching high school English classes.

Another time J told us about another patrol he was on where it was discovered that a large North Vietnamese force was headed right their way. He got on the radio and called in artillery and air strikes and was ordered to get his platoon out of there as fast as possible. They started hauling ass and had gone a few kilometers when it became totally apparent that one of his Marines was constantly falling back behind the rest of the unit. He halted the platoon and his Marines immediately went into 360 degree defense perimeter mode. He went to the man who was straggling. It turned out that this guy was a recent replacement in the platoon who had screwed up some rear-echelon job back in Da Nang and was sent to the field as punishment. This was, unfortunately, an oft repeated ploy by rear echelon junior officers to rid themselves of marginal Marines. Thanks guys. Your junior leadership incompetence has now placed a front-line platoon in danger of being wiped out because of a Marine who never should have been out on the front lines in the first place.

Think about it! Rear echelon officers were punishing marginal Marines by sending them to front line units whose very existence depended on being loaded with exemplary, motivated Marines who would lay down their lives for their friends.

Well the kid was overweight and out of shape and that's why he couldn't keep up. J squatted down in front of him and pulled out his service .45, flipped off the safety, pulled back the slide and chambered a round. He placed his M1911.A1 Colt .45 Semi-Automatic pistol to the side of the kid's head and said, "Now look here, son, this platoon is at risk of being overrun because you can't cut it. You have choices: you can either get up off your ass and start running with the platoon, or I can put a bullet in your brain." The kid found his motivation and took off. J said that the kid turned out to be a really good Marine after that.

J told me that what disturbed him the most was that he really would have pulled the trigger.

This we ask of our junior officers in service to their country.

So how does that compare with your job?

Within two months of my arrival at Geiger, one Friday afternoon a message came from 2nd Division Headquarters that I was to report to 2nd Division Headquarters no later than 1600 (4:00) that same day to be assigned to a six-month Mediterranean deployment which was departing Saturday morning. 2nd Marine Division was the principle tenant of Camp LeJeune. As such it was responsible for deploying reinforced battalions of approximately 1200 combat trained Marines. These Task Forces were called Marine Amphibious Units, or MAUs. Today they are called Marine Expeditionary Units – MEUs. They were deployed, at that time, every six months to the Mediterranean, and every four months to the Caribbean. Today we no longer deploy MEUs to the Caribbean.

It turned out that whoever had been assigned as the disbursing officer for the current Mediterranean deployment had become unavailable at the last minute (broke his leg or something).

Dean Lucht did not want to see me go so soon for two reasons: one, he needed me there as his assistant disbursing officer, and two, he was sympathetic to the fact that I had just come back from a year away from my bride. He told me to get out of the office for the rest of the day and he would send a message back that I was away for the weekend on a 72-hour pass.

That saved my bacon for sure. I knew being assigned to LeJeune meant that deployment would be in my future, but I really wanted a little break before I deployed.

It was an interesting time for the Marine Corps. Much like at Iwakuni, both the peace movement and racial unrest were a constant presence. I was standing duty as Officer of the Day at the main base one steamy Saturday night in July of 1971 when at about 11:00 PM the Sergeant of the Guard came in to tell me that there was a large gathering of black Marines congregating under a tree over near the PX. The commanding general of the base had issued orders that gatherings of four or more Marines anywhere on base after sunset was forbidden. This order had been issued because of the fear of racial unrest.

I told the well-intentioned sergeant that I would handle it. I stood up, unbuckled my duty belt, disconnected the .45 caliber pistol and holster, rebuckled my belt, and gave the Desk Sergeant my holster and pistol. He said, "Lieutenant, are you nuts?" I said, "No, I'm just being practical."

I went out to see what was going on wearing just my basic uniform, my olive green duty belt, and my red armband with gold OOD lettering to represent my authority. I figured that if things got ugly my pistol would be useless. In the best of circumstances, it would be stolen. At the worst, it would be used against me.

I walked up to the group and said, "Good evening, Marines. What's going on?" They were surprised to see me, the OOD, there alone, white, and unarmed. "Where's your posse, Lieutenant?" one of them asked. I said that my posse needed some rest, and I thought I'd see for myself what was going on. We ended up actually having a good conversation, particularly when one of them said that they were discussing their feelings about the discrimination they experienced in the Corps. I told them that I was sympathetic, that I was aware of the systemic discrimination in job advancement and promotion opportunities, but there was not much I could do about that at midnight on a hot summer Saturday evening. And that there wasn't much they could do at that time either. I told them that I would pass their concerns along to the command. Then I told them it was time to break up the party and go back to their barracks, and they dispersed.

When I got back to the guard shack, the Sergeant of the Guard told me (with all due respect) that I was crazy to do that and I should have called out the reaction platoon to break up the gathering. I told him that that was the worst thing to do. I said that in the future that he should handle such situations respectfully and that the situation would generally end peaceably.

I wrote the incident up in the logbook, and the next morning the Battalion XO (a Marine major and Naval Academy man) read it when he was relieving me. As I stood there at parade rest in front of his desk, I saw the surprise on his face when he got to the damning entry. He, of course, missed the point. Did I mention that he was Naval Academy? (What the hell do they teach these guys there??!!) He berated me for leaving my pistol with the Sergeant of the Guard. He said that I should have called out the reaction platoon. I said something to the effect of yes, sir, I have learned my lesson, and then went back to work.

I was rapidly coming to the conclusion that the officer corps suffered from a hardening of the arteries, a severe lack of creative imagination, and a paralyzing fear of change. That conclusion was reaffirmed many times before I retired.

The daily routine at LeJeune, i.e., Camp Geiger, was not very difficult. I had a pay section with SSgt. Netzle (E-6) in charge, and I had a sergeant (E-5), a corporal (E-4), and two lance corporals (E-3s) as clerks, as well as two women civilian clerks working for me. Kathy usually drove me to work and picked me up every day. (Kathy, by the way, would spend the academic year of 1971 to 1972 teaching math at Jacksonville High School. Through her teaching experience we made several civilian friends that we kept in touch with at Christmas for many years.)

I always packed my own lunch, and after I had run three miles every noontime I would shower and eat my lunch at my desk. There were shower facilities available to me near the office. Fortunately I had J to run with every day. On Fridays J and I would drive over to the main base to run with a man who was both one of his old friends and one of his professional mentors. The officer was an older Mustang (former enlisted, in case you forgot) Major who had studied not only the science of running, but also the science of improving our ability to do pull-ups for the PFT (Physical Fitness Test). He had come up with some excellent training techniques to improve our strength, stamina and speed.

Life was, and I understand still is, miserable for single Marines at Camp LeJeune. By the time Wednesday afternoon rolled around, so the legend goes, Marines at LeJeune could be seen walking around tilted north toward Washington DC. Every Friday from noon onward, large groups of Marines who had been granted a 72 hour or 96 hour liberty gathered at 'Swoop Circle,' a parking area near the circle at the center of the base where the American flag flew from a tall pole. Most of the hopeful travelers had hand-made signs that said, usually, 'Washington' or 'New York City.' Anyone who had a car, or for the more enthusiastic entrepreneur, a van, and was going to either of these places drove into the circle and picked up anywhere from two to seven passengers who were willing to pay the going rate to get to either of these places (in 1971 the going rate was $10 to DC or $20 to NYC each way). The driver would drop them at some central location in the designated city and pick them up at the same location the following Sunday. This was how LeJeune emptied every Friday, and how it got refilled every Sunday night.

After I transferred to the main base at LeJeune, I discovered that I had an enterprising Sergeant Ramos working for me who was a New York City native and owned a nine-passenger van. He charged his passengers what was then considered a fair rate of $40 for a round trip to New York City. His entrepreneurial efforts grossed him $360 every weekend, more than enough to cover gas and maintenance expenses, as well as a tidy profit. Keep in mind that gasoline in 1971 was still only about thirty cents per gallon.

Before that, over at my own little enclave at Camp Geiger, my troops also did the same thing. SSgt. Netzle was from Milwaukee and owned an enormous '59 Pontiac Bonneville four-door sedan. It was the size of a small aircraft carrier. Every time he could get a 72 or 96 hour pass, he would load it up with six Marines from the mid-west and leave at noon Friday for Milwaukee. He would get there at noon Saturday, party all night, then Sunday (after church, no doubt) load up his passengers and return in time for work Monday morning. He always brought back kielbasa, a Polish sausage that is unique to the Chicago-Milwaukee area.

On weekends Kathy and I either had people over for dinner or we went out to eat. One of our favorite dining establishments was the Country Squire located about forty-five miles from LeJeune. The food was fabulous, and in keeping with the rules of North Carolina, it was a "dry" establishment. If you wanted to drink, you brought your own beer or wine. If you wanted a mixed drink, then bring your hard stuff and the restaurant would provide the ice, club soda, and whatever other mixers you required.

Despite the logistics of remote dining, we often went to the Country Squire.

Quite often during the day on Saturday and Sunday in the extended North Carolina summer, we went to the beach. We usually went to West Onslow beach, which lay on the coast just south of the base. It was beautiful, and quite often nearly deserted, even on weekends. The ocean there was fabulous and we did a lot of body surfing, many times seeing dolphins frolicking off shore. Today the whole coast of North Carolina is covered with condos. It just ain't the same.

One Saturday we returned from the beach around three o'clock, and we weren't in the door a few minutes when the phone rang. It was Capt. Lucht telling me he had been trying to reach me all day and that I had better get to the office right away as someone had tried to break into my safe. I immediately put on a uniform and got over there. The place was swarming with MPs and investigators from the Naval Investigative Service (NIS, later to become NCIS – Naval Criminal Investigative Service – and the source of a long running television series).

There in our disbursing office, in the middle of the floor, was an acetylene tank with hose and torch. Apparently two of my clerks (the two lance corporals) had first broken into the office, then tried to cut open my safe. Failing at that they tried to open Capt. Lucht's safe with the acetylene torch. Fortunately they did not know how to use the equipment properly and failed in their attempt. Their criminal success hinged on one Junior High School Shop course (on the use of the acetylene torch) they never took. One of the lance corporals was married. His wife was an accomplice and get-away driver.

I told the NCIS investigators that during the past several weeks I had noticed these two hovering around my desk apparently trying to see the combination as I opened the safe every day. I thought they were kidding around and had not thought much about it. The interesting thing is that my safe only held about $10,000 in cash, and Capt. Lucht's safe held maybe three times that much. It wasn't much money, all things considered. The office break-in was discovered by a passing MP who saw the door ajar. The perpetrators had already left in frustration and were now on the run from federal authorities. Within a month they were caught and admitted their guilt. The Marines were sentenced to six years hard labor, six years forfeiture of all pay and allowances, and a Dishonorable Discharge, or as we affectionately referred to it, 'Six-Six and a Kick.' The Lance Corporal's wife got a similar sentence in civilian federal court as she was considered just as culpable as they were. Now that's something they can tell their grandchildren.

In late August of 1971 my time had come and I was transferred to the main base 2nd Division disbursing office to await assignment as disbursing officer for a deployment. I now would be working directly for Major John English, which turned out to be a very pleasurable experience, just as I had experienced with Dean Lucht. Both of these men were excellent leaders who put their people first, and, as a consequence, were able to easily complete their missions.

Once settled into my new assignment, I found myself in charge of a pay section that handled the accounts for 2nd Battalion Sixth Marines, commonly referred to as 'Two-Six, or 2/6'. I had Corporal Clark (who did a great Elvis impression), Lance Corporal Moad, and Private First Class Jensen working for me. They were all good Marines and excellent disbursers.

2nd Battalion, Sixth Marine Regiment was ramping up for a four-month deployment to the Caribbean that would include 3 weeks at Jungle Warfare School in Panama and several weeks of amphibious training assaults on the island of Vieques, just off the tip of Puerto Rico. Over the next few months, orders were cut for units to be attached to 2/6. Such units consisted of tanks, artillery, engineers, Recon Marines, ANGLICO (Air Naval Gunfire Liaison Officer), helicopter aviators (and their support units), MPs, Mail Clerks, and, of course, the Disbursing detachment. The addition of each of these units increased the responsibilities of my pay clerks. I'm proud to say that my guys rose to the challenge.

As the Disbursing Officer for the battalion, I was assigned the additional duty of being the officer-in-charge of the various attachment units: MPs, Mail Clerks, and assorted other groups that otherwise had no "parent." The logistics of such a "mount-out" are enormous. The battalion had to coordinate not only the movement of the infantry units and their equipment, but also the tanks, artillery, engineers, communicators, administrators, aviators, and also the Disbursing Officer and his entourage.

In the past, apparently, other disbursing officers had simply acquiesced to the group think and joined the battalion convoys to Moorhead City. Major English told me to check the procedures outlined in the Navy Comptroller's Manual (the Disbursing Officers' Bible), and then make the right decision.

After checking the NAVCOMPTMANUAL, I naturally sought the advice of those who had gone before regarding how the shipment of the disbursing team and more than $300,000 in cash was to be handled. And I reached a conclusion. I informed the 2/6 battalion commander, Lt. Col. Enoch (who happened to be a graduate of Mount St. Mary's College where I had initially enrolled), that, in accordance with the NAVCOMPTMANUAL, the disbursing team and its money had to be convoyed over to Morehead City (our debarkation point, about 40 miles north of Camp LeJeune), separately, and unscheduled from the rest of the convoy. He wasn't happy about that, but I said it was for security of the cash and in accordance with Navy regulations.

In the new travel plan that I implemented, the MPs provided me with a separate convoy of three trucks and twelve guards, each armed with 12-guage pump shotguns.

I went to the bank on base with a U.S. Government check made out in my name in the amount of $330,000 dollars (a copy of which can be found in one of my photo albums). It was a heady moment to collect such a sum, count it out, and stash it in a small government safe.

With the advice of the senior Military Police NCO, I ensured that the front and rear trucks were loaded with fully armed MPs wearing bullet-proof vests, and the middle truck of our mini-convoy was loaded with me, my three clerks, our personal gear, three mail clerks, and a safe containing more than $300,000 in cash.

We conducted a high-speed transit from LeJeune to Morehead City and arrived safely with sirens going, blue lights flashing, and, as an additional flourish, the MPs jumping out of their vehicles facing outward in a 360° perimeter with their 12-guage pump guns at the ready. We off-loaded the safe containing the cash and brought it up the gangway into our ship, the USS Shreveport, to be secured in our office afloat.

Observing this from the Bridge of the ship, not only was the battalion commander impressed, but the Naval Amphibious Squadron Commander (the Navy Captain in charge of our 4-ship squadron), who also was designated Commodore for our deployment, was also awed by our display of force.

Having secured our office, and having ensured that my troops were properly billeted, I dismissed our MP escort with great thanks. The remaining MPs who were going on our deployment settled in to the shipboard routine. The following morning, bright and early, our ships pulled out into the Atlantic and pointed their bows to the south.

Our amphibious squadron was made up of the USS Shreveport and the USS Nashville, sister ships both designated LPDs (Landing Platform Dock). These ships had characteristic traditional Naval ships' bows and mid-sections, but also were characterized by a flat rear deck able to land two helicopters simultaneously. Below that deck was the well-deck holding the vehicles we would need in our landing: jeeps with trailers, small trucks with trailers, and large trucks with trailers which were stowed in preparation for deployment. We also had the USS Portland, an LST, which stood for Landing Ship Tank that held the AMTRAKS, the tracked amphibious vehicles that would carry infantry Marines to the beach. We also had one other ship (the name of which I can't remember), designated LSD for Landing Ship Dock that carried the tanks, artillery, and small landing craft and amphibious tractors (AMTRACs) that would go to each ship, load up with Marines, and then head to shore. The Shreveport and Nashville carried mostly infantry Marines. The tailgates of those two LPDs could be lowered to the horizontal and the well-deck flooded allowing landing craft and AMTRACs to drive in and load up with Marines in preparation for an amphibious assault on a beach.

I had never been on the ocean out of sight of land before. As a kid boating in the waters off Norwalk, you could almost walk across Long Island Sound as it was only three miles across from the southern Connecticut shore. As a Marine, all my previous experience with amphibious landings had been in training exercises within a few miles of shore with the land plainly in view. On this, my first deployment, I immediately fell in love with going out on deck at night to see the stars so incredibly clear with no light pollution from land. I also enjoyed looking down at our wake to see the water glow from the churned up phosphorous in the water. During the day on our journey south, we were often accompanied by schools of flying fish.

Shipboard life for Marines can be pretty boring. Our main job was to keep out of the way of the sailors who were doing their jobs. I made it a point to spend a lot of time in the office with the clerks who were preparing the pay records for pay day. The Headquarters company admin officer, Marine Captain

Wells, also suggested that while afloat it was a good time to conduct a complete audit of pay records against the personnel records of all 1200 men deployed in the reinforced battalion. This made sense, as we were in a somewhat hermetic environment. This was way before the internet and instant electronic communications. Everything was done manually, and as we were a self-contained organization afloat, we could conduct a detailed audit of our pay and admin systems.

We began the audit with the 350 Marines on the Shreveport. We planned to audit the other Marine records when we were all ashore together. Ultimately, as a result of the on-board audit, we were able to reduce pay errors for those on our ship to less than 1 percent by the next payday. The exercise also kept my clerks busy and away from the other Marines in the battalion who spent so much time every day in endless poker games and other forms of gambling. All illegal, mind you, but tolerated.

At night our squadron would travel south in zig-zag patterns. I was told that the Russian submarines patrolling off our coast in international waters would try to track our movements. I learned that each ship's propellers had a unique sound signature. No two ships sounded alike. The Russians would try to identify the sound signature to confirm the identity of the ship. The US Navy's job was to make that difficult. On board the Shreveport, the ship's captain had some of the sailors form a rock band, and when SONAR detected a submarine in the area, the sailors would be alerted and quickly go down into the well-deck, plug in their instruments and play for several hours to mask the sound of the propellers.

One of the duties of every Marine officer is to make sure that the troops are getting fed properly. That would mean a regular check of the mess deck to be sure they were getting a sufficient quantity and quality of food. The officers were fed separately in the Ward Room up in the O-2 Level (that is, two decks above the main deck) in what was always referred to as 'Officers Country.' For the officers, the system was that each officer in the mess paid a pro-rata share to cover the expense of the meals. The Navy Supply Officer handled the payments. If we were going to be aboard for 21 days, we only paid for 21 out of 31 shares (31 being the standard month unit). Therefore, when we were ashore, we were not billed for our meals.

The problem was that we Marine officers all felt that we were paying a little bit too much for our meals, given that their quality was nothing exceptional, just meat and potatoes for dinner as well as hot dogs a little too often, but we didn't complain to the Navy, at least not just yet. That was to come later.

I had made good friends with a number of officers from 2/6, including Navy Lieutenants Ed Dolan (the dentist) and Rich Goldhammer (the physician), and several other of the Marine Captains and Lieutenants aboard the Shreveport including Lt. Joe Gould (Supply), and Lt. Covington P. Stanwick (Air-Naval Gunfire Liaison Officer). Ed Dolan, it turned out, had graduated from Mount St. Mary's in 1967. He was an avid golfer and promised to give me lessons when we got back to LeJeune.

After a week or so, we arrived on the east coast of Panama, just above the locks, and debarked all the troops to spend three glorious, fun-filled weeks at Fort Sherman, the Army's Jungle Warfare School.

Once we were ashore, the ships proceeded down to Cartegena, Columbia for more Naval maneuvers and replenishment. While the rest of the troops went to the jungle for training, Disbursing and Administration, along with MPs and Mail Clerks, set up shop in the main base headquarters building at Fort Sherman. Our Supply Officer, Joe Gould, remained aboard ship with the Navy to arrange resupply of the Marines following our Jungle Warfare Training.

I moved my troops into an ante-room of the Army Commanding Officer's office area where the Army Sergeant Major got us settled in. I chained our safe to a water pipe and the clerks set about getting ready for payday three weeks hence when the rest of the Marines finished jungle training. My guys worked diligently with the admin clerks from the rifle companies (who had also remained in the rear) to reconcile pay records with the daily Unit Diary entries into personnel records in an attempt to reduce pay day errors. These efforts at producing a solid audit were paying off.

The Unit Diary was a daily administrative accounting of promotions, legal actions resulting reductions in grade or fines, and other miscelania. Disbursing based its actions on what was promulgated in the Unit Diaries of each company or attachment. If an operational unit said So-and-So was promoted to the rank of (blank), then we entered the transaction onto the Pay Record and paid accordingly.

Occasionally I needed a mental health break. One of the more pleasant things I did at Fort Sherman was to take advantage of their zoo. The U.S. Army had created a wonderful zoo showing all the flora and fauna that one could experience in the Panamanian jungle. I had the opportunity to pet a puma (I have pictures) as well as hold a young boa constrictor. The boa (just a baby as army SSgt. Lindor explained) was six-feet long and was more than a handful to deal with. Again, I have pictures.

SSgt. Lindor regaled me with the story of an arrogant Army colonel and his wife who were stationed elsewhere in the Canal Zone. One Sunday they showed up at Fort Sherman and demanded a tour of the zoo. SSgt. Lindor was all too happy to comply with the request but balked when the colonel's wife was leading a toy poodle on a leash as they approached the entrance to the zoo. He said, "Colonel, I'm happy to give you a tour of the zoo, but the dog will have to stay in your car as its presence will upset the other animals."

"Nonsense," said the colonel. Fifi is a member of the family and she goes wherever we do."

"All right, Colonel, but I'm not responsible if something goes wrong."

So the zoo tour began. When they got to the alligator pit, the colonel's wife put Fifi up on the retaining wall and started to lead the dog around the pit. Suddenly Fifi lost her balance and slid down the concrete retaining walls. Two alligators promptly chomped Fifi into doggie heaven. SSgt. Lindor simply said, "Adios, Fifi." And walked away.

The Army colonel tried to shift the blame for the occurrence to SSgt. Lindor, but fortunately for the Army, no one was buying the story.

By the third day at Fort Sherman I was bored out of my mind. I had more than enjoyed my time talking with the Sergeant Major. He was a Green Beret who had been on the famous Son Tay raid into North Vietnam. Army Green Beret forces had rehearsed the raid for months at Fort Bragg, North Carolina, on an exact mock-up of the Son Tay prisoner of war camp. The mock-up was the result of high-altitude photographs taken by our SR-71 supersonic spy plane that travelled at 80,000 feet and at speeds exceeding Mach 4.

This highly trained and motivated raid unit composed of elite army soldiers after months of exhausting training, flew over to Vietnam and launched the POW rescue raid at night. The raid went like clockwork, except for one fact: the prisoners had been moved out of Son Tay just the day before the raid. Of such is the fog of war.

Being bored as I was, I asked the Sgt. Maj. If he would look after my troops while I went out to the field and took part in the training. He said sure, and I hitched a ride to the jungle training site. It was quite an experience. When I arrived in the bush the other officers looked at me like I had grown a third nostril. What the hell was the Disbursing Officer doing out here when he could be back in a clean environment and getting a hot shower every day. I assured them that I was where I wanted to be, so the S-3 (battalion operations officer) sent me to join E Company and share their training experiences as a shadow to their CO.

E (or Echo in the military alphabet) company was commanded by a 1st Lieutenant "Rod" Rodriguez. Rod was one of the best officers I ever saw. Like me he was a former enlisted guy and you could see the respect that he commanded from the lowest private right up to his First Sergeant. He had come from a hard-working Mexican immigrant family in Texas and had struggled to overcome his poor background. He had one year of college before enlisting in the Marine Corps at age 19 and, like me, shortly after Boot Camp qualified for the Enlisted Commissioning Program. I took to him immediately. He was charismatic, funny, as handsome as a movie star, and really good at being a Marine infantry officer. Years later I found out that he had a great career and earned promotion to Brigadier General before he retired. In my experience with Rod in Jungle Warfare School, I was inspired by many of his leadership traits, and as my career continued, I drew on those experiences.

It was a pleasure to be sharing the Echo Company training experience, and I may be the only disbursing officer in history who got a certificate of completion for the Jungle Warfare Course. Rear echelon officers customarily did not go out into the jungle.

Toward the end of the training, Col. Enoch asked that payday be held on the day they returned from the field, so I went back two days early and got the crew cranked up for pay day.

While I was gone, my clerks had gotten together with the various admin clerks from the battalion and were continuing the audit of pay record against Unit Diaries and Service Record Books to ensure that there were no discrepancies. Pay day came off without a hitch and everyone was pleased.

Col. Enoch also said that he would accept nominations for meritorious promotion to the rank of Corporal for any Marine who was eligible. LCpl. Moad had done an outstanding job. He had the heaviest load of all my clerks handling more than 500 pay records (of the 1200 total) himself. He was meticulous in his attention to duty. I wrote up a glowing recommendation for his promotion and submitted it. The colonel had assembled a board consisting of his Executive Officer (a Naval Academy drone if ever there was one), and the commanding officers of each of the infantry companies and the CO of the Headquarters Company to interview each of the candidates. They would accept 10 for meritorious promotion.

Moad went into the board looking sharp, but when he came out, he was dubious of his chances. I asked him why and he said that they only asked him infantry related questions. Later on Captain Wells, the Headquarters CO told me that Moad had been sandbagged. One of the captains on the board made a big issue out of making Moad look bad by throwing complex infantry related questions at him and not letting him talk about his competency as a disbursing clerk. Wells said that the captain had a reputation for being a jerk, but that he was the son-in-law of the Camp LeJeune commanding general. The XO and most of the other officers caved under his pressure, and it was only Wells and Rodriguez that supported Moad's promotion. They were outvoted.

Moad appreciated what I had tried to do for him, and I was able to get him meritoriously promoted when we got back to LeJeune. To his credit, he never slacked off in his work after such a let down. That, to me, was a real mark of maturity.

The troops were given liberty in Colon, Panama. I had driven through Colon shortly after we had arrived, and I had never seen, nor could I ever have conceived, nor will I ever even attempt to describe such poverty. So the Marines of 2/6, minus me and my clerks, after being cooped up in the ships for weeks, then pushed through the rigors of jungle warfare training for three weeks, let it all hang out hitting every nasty bar in Colon and spending their pay the way Marines do. I will not elaborate, but for the next several weeks, Dr. Rich Goldhammer was using up his supply of penicillin.

We debarked from Panama a few days later and headed for San Juan, Puerto Rico, for four days of liberty. En route we ran into a tropical storm that had almost everyone except the ship's crew lying low in their bunks. But motion sickness never bothered me, so I got up, ate breakfast, and had a cup of coffee. Then I went out on deck to enjoy the storm. But I always remembered a bit of good advice: when you start to feel a bit nauseous, get out on deck, focus on the horizon (a fixed point), and adjust your gait to the motion of the ship. I only had to do that a couple of times. We finally broke through the storm into calmer waters and headed for Puerto Rico.

An interesting development occurred during that time. The Supply Officer had not left the ship in Panama but had continued on with the ship to Cartegena to arrange for a re-supply for the battalion. He told us that as soon as we left the ship, the officers' meals had improved. They were having steak, lobster, or shrimp every night. We were pissed, but we weren't sure how to handle it as we were at the Navy's mercy for at least another month.

We arrived in San Juan and docked right along the main avenue. If you debarked the ship and turned left, you headed for 'Old San Juan,' which was the historical area. Just cross the boulevard and you were also just a short walk from the beginning of the luxury hotel strip.

After getting the troops off around noontime for liberty, the officers were granted liberty around 1500. About 15 of us lieutenants and captains took a short walk straight across the boulevard to the Caribbe Hilton, the first luxury hotel along the strip. We walked through the lobby and down a few steps to a huge rectangular bar overlooking the pool and the ocean, and there we set up shop until dinner time.

As we were sitting there, periodically an absolutely stunning looking woman in a long evening gown would come from the lobby to the top of the stairs overlooking the bar. She would smile at us and walk back into the lobby. Upon her third arrival at the top of the stairway, I led the gang in breaking into the song, "A Pretty Girl Is Like A Melody" and she just laughed and went back to the lobby. Captain Wells decided that he had to find out just what her story was. Off he went to the lobby with our best wishes, and he returned minutes later with a stunned look on his face. He said he asked her if she was waiting for someone, and she replied, "My rate is $50 per half hour."

So much for a pretty girl.

After a few beers it was getting along toward dinner time, so we walked back across the boulevard and went to the Navy Officers Club right along the harbor. We had a fantastic meal, and there, entertaining the diners, was a small Mariachi Band of three guitarists and a bass player doing bolero (serenade) songs. They moved from table to table, singing beautiful Spanish songs. I asked them to play '*Sin Ti*' and '*Malaguena Salerosa*,'. They did and it brought the restaurant to a standstill. They did it so beautifully. I tipped them well. Rod told me I had done good.

The rest of liberty went by with doing tourist things in the daytime and drinking from one luxury hotel to another at night. We wrapped up our four-day stay in San Juan and then embarked aboard ship for Roosevelt Roads on the opposite side of the island. Rosy Roads was a Naval installation (now closed) where the ships were to be refueled and replenished, so we had an overnight stay there.

The usual crowd – about a dozen of us – left the ship and went up to the "O" Club for burgers that night, and, after a more thorough briefing by our Supply Officer, over several beers, became more enraged at how we had been treated in the ship's officers mess. We reflected on our Supply Officer's report that the Navy officers dined like kings when the Marines were ashore. We had been hosed. Since our re-embarkation aboard ship, our dinner fare had returned to hot dogs, cold-cut sandwiches and the like. We were pissed that we were subsidizing high quality meals for the Navy officers but were only eating common fare for ourselves.

So around midnight, under the direction of Captain Wells, the Headquarters Company Commander, we formed up in a column of twos, double-timed (sloppily) back to the ship, up the gangway past the officer of the deck, who raised futile protests about our failure to follow ship-boarding protocol, and into the Wardroom, where we broke open the locked refrigerators and prepared ourselves a feast. Because we were officers, the Wardroom stewards didn't know what to do, but one of them finally woke up the ship's XO who came down to find out what all the fuss was about.

There we were, the officers of 2/6, pissed and gorging on the great food that we had been denied.

Wells told the XO we were extremely outraged and that we had been charged full rates for second rate meals while the Navy officers ate like kings after we got off ship, and that we were not going to subsidize that kind of situation without a fight. The XO then told us to get out and he would look into it. Reluctantly wiping the crumbs from our mouths, we left. The next morning we were all standing tall in front of the ship's commanding officer, a Navy Captain, a person who at sea and within full naval tradition had the powers of a god. With Col. Enoch looking on us like he would gladly put us all in front of a firing squad, we made our case, or at least Captain Wells made it for us. The CO chewed us out for breaking open the mess larder, but he also assured us that we would not be cheated again. He subsequently relieved the Navy supply officer and sent him home.

Surprise, surprise, as Gomer Pyle once said. Our meals improved dramatically.

Next stop was Vieques, an island off the southeast coast of Puerto Rico. Vieques was an interesting island. One end had a small village and a long, curving, stunningly beautiful beach and several exclusive mansions that had been built by French plantation owners for their mistresses. These classic dwellings were now small, intimate luxury hotels. The other end of the island was part Naval gunfire range and part amphibious landing and infantry training area. (It was, sensibly, subsequently shut down and cleaned up in the 1990s.)

We made a tactical landing on the beach, and being as about as unessential to the tactical operation as we could be, Ed Dolan and his dental assistants, the mail clerks, and the disbursing crew were in the final (I think fifth) wave of the landing. My crew and I muscled the safe off the landing craft and on to the beach along with all of our gear. Vieques was to be our home for the next three weeks.

While we were standing on the beach awaiting a truck that would bring us up the hill to the company headquarters area, Ed Dolan reached into his seabag, pulled out a sand wedge and several golf balls, and proceeded to work on his golf bunker swing. It was one of the more absurd moments I ever experienced in the Marine Corps.

Finally our transportation arrived and brought us about a mile and a-half inland to the top of a hill where several Quonset huts awaited our occupation. We grabbed one end of a long Quonset hut for our disbursing office and set up shop. On the opposite side of the wall from us, the battalion headquarters had set up operations. Across the 'courtyard', i.e., a hard-packed dirt yard, from our office was the officers' club, another Quonset hut on a concrete slab much like the club I had been in on Okinawa. We Marines really know how to be posh.

Our living quarters were, to say the least, modest. All the non-grunt junior officers (Supply, Intelligence, Artillery, Air-Naval Gunfire Liaison, Administration, Disbursing, Dental, Medical, et al) were in one open, decrepit, Quonset Hut barracks. Unlike Boot Camp the bunks were single and not double-stacked. The mattresses were of dubious merit and sanitary quality.

From my limited experience, I knew that Naval officers who had never served with Marines were elitist to the core. They would have blanched at the idea of rooming in a shabby, beat-up tenement where we had been assigned.

Our new home had single beds from one end of the building to the other, and no privacy. Nonetheless, we non-infantry Marines (the infantry were out in the field in tents) were content that we were out of the elements. All of us got along together and had some pretty good laughs along the way.

Most of the laughs during our sojourn were provided by Lt. Covington P. Stanwick, heir to the Stanwick Corporation fortune. Covvy was a complete ham. One night he regaled us with his rendition of West Side Story, playing all the major parts himself. Another time he did all of Sesu Hayakawa's lines as commander of the prisoner of war camp in 'The Bridge On the River Kwai.' He was a natural mimic and kept us in stitches when he wasn't out in the field coordinating Naval gunfire and airstrikes.

After a couple of days on Vieques, Corporal Clark approached me with a proposition. He wanted to move himself and the other two clerks out of the barracks they were in and move into the office. He said that all the admin and support people housed in the barracks were going into the town of Vieques every night, then coming back to the barracks trashed and raising hell all night. Nobody could get a decent night's sleep. So I granted them permission to make the move. That evening he and my two other clerks brought their mattresses into the office. During the day the mattresses were stacked against a wall. At night the clerks would push the field desks to one wall and then lay out the mattresses.

It soon became apparent that the hot afternoons on the island were not conducive to administrative office productivity, so I proposed to my people that if I could get permission from the CO, I would like to rearrange the work schedule. My plan was to work from 0730 until noon, lock the shop, and then take a run to the beach (about a mile and a-half). There we would have lunch courtesy of the medical staff who would bring sandwiches from the mess hall to us in their ambulance. That seems excessive, but the medical folks were headed to the beach every afternoon anyway.

The plan was to spend the afternoon swimming, have some fun, then run the mile and-a-half back up to the office, shower, go back to work around 1600 (4:00), and work into the evening breaking only for evening chow.

The bottom line was that we were doing a three mile run every day in a combat training environment.

The troops all loved the idea, but I told them that everything would be conditional on their producing a completely accurate payday on the demand of the CO. They assured me they could do it.

I went to Col. Enoch and proposed my plan. Just as when we were in Panama, pay days were not the regular twice a month occurrence that they were back at Camp LeJeune. Operational requirements required a flexible payday schedule. Pay day would occur when the colonel wanted it.

He looked at me and said that it was highly unusual, but as long as he got his pay day when he asked for it, it was okay with him.

Thus began our unusual schedule of work mixed with physical training. It worked out great, and the medical staff agreed to bring our lunches. The morale of my guys went sky high as they didn't have to put up with the barracks nonsense and they got to go to the beach every day. As a result, they actually worked longer days, many times up to midnight.

Second Battalion Sixth Marines was actually doing combined training with a battalion of 800 Brazilian Marines. The Brazilian Marines had a US Marine Colonel as their advisor. One night after a tough training day, Col. Enoch and several members of our battalion staff were in the club having a beer when the Marine-Brazilian liaison colonel joined us. We were talking about this and that when Col. Enoch mentioned my unusual training schedule. The other colonel found it hard to believe that my guys were self-motivated enough to be putting in a lot of extra time at work on their own. It was late in the evening and getting along toward midnight, so I got off my bar stool, went to the door and saw across the 'courtyard' that lights were still on in the office. I turned and said to the Colonel, "Sir, I bet you $5 that if you walk across the courtyard right now you will find my guys still working."

He took me up on the bet and we walked across the way. I pulled open the screen door and called out, "Attention on deck!" and my troops, who were all sitting at their desks and working on payroll, popped to attention. The colonel said, "I see it, and I still can't believe it. As you were, Marines!"

He took some time and asked each one of them what they were working on (which impressed me no end) and came away dazzled. I knew my guys would rise to the occasion.

We finally wound down operations, re-embarked aboard ship, and sailed for home. It was now mid-May 1972 and I was scheduled to leave the Marine Corps on July 12.

When we got back to LeJeune I was assigned an additional clerk, a LCpl. McDonough, who was a really sharp and squared-away kid. He helped lift the burden that my three clerks had been carrying for the past four months and played a critical role in doing the final Pay Record/Service Record audits for the entire battalion. On the first pay day back after we got off ship, we had a problem. McDonough came up to me and said that there was a really pissed off Captain at the counter with one of his troops who apparently had been paid incorrectly. So I went up to the counter, and low and behold, it was the nasty Captain who had sunk Moad's chances of getting meritoriously promoted – the son-in-law of the commanding general.

I asked him politely what the problem was and he proceeded to chew me out in front of the entire office. I kept my cool and said I would find out what went wrong. In the meantime, McDonough was on the phone with the Captain's admin clerk and found the source of the problem. It seemed that the young Marine had been promoted, but his admin office had neglected to tell disbursing. We informed the officer that the problem had been on his end at the company administrative level, but that we would be glad to correct it. He did not seem too happy that the problem originated at his own company.

Let me clarify something here. This was a common admin glitch, and one that was easily fixed. The problem was that we had too many officers who were willing to throw their authority around without first looking for the source of the problem.

So we fixed the problem, paid the difference, and I thought that was the end of that.

It turned out that McDonough and the rest of the office were really pissed that the captain had tried to humiliate me in public, and general's son-in-law or not, something was going to be done.

A year later, as a Training Assistant on active duty orders for the summer, I was back at Camp LeJeune between my junior and senior years at Mount St. Mary's College providing liaison assistance between Reserve and Active Duty units at Camp LeJeune.

Hoping to see a few old friends, I stopped in to the 2nd Division Disbursing office to say hello to everyone. The young trooper who had joined our unit just before I went off active duty, LCpl. McDonough, was now a corporal. Shortly before I arrived he had received orders to go to OCS in the fall under the Enlisted Commissioning Program (my own alma mater). I congratulated him on his accomplishment. He thanked me and said, "Hey Captain (I had been promoted), do you remember that nasty captain from 2/6?" I said, "How can I forget?"

McDonough then said, "Well we all felt bad about you having to go through that humiliating crap with him. We didn't want you to get in any trouble, so we waited until a few weeks after you left active duty and then mailed his pay record to Marine Barracks, Adak, Alaska.

I must explain. Under the old pre-computerized pay system, it was not without precedent to do something like this. It was the only way for a low-level clerk to gain any sense of justice over an abusive superior. It was an unwritten rule among disbursers that if you received a pay record in the mail that had no documentation with it, then it was a revenge job, and the only action you took was to tape it to the underside of your middle desk drawer until it either rotted or the Rapture came.

It would be a felony to destroy a pay record, but losing one deliberately was hard to prove. In such cases, the only way for the victim to get paid was to open a new, temporary pay record. The best part was that with a new pay record, the bastard could only be paid at the lowest rate of pay for his rank without his every-two-year longevity increases. As an added bonus, his living quarters and food allowance were also cancelled. Also, if the victim had any allotments going out for car payments, savings plan, or mortgage payments and the like, those were terminated also. After six months, if the original pay record didn't show up, Disbursing would open a new Temporary-General pay record and restore some back pay and reinstate the allotments without the longevity increases. After a year, full restitution would be awarded. So the bastard would eventually get his money back, but in the meantime he would suffer. In most cases, that was enough.

"Boy was that captain really pissed now," said McDonough. "We told him what the process was and that he would have to wait a year before everything was restored. He was not a happy camper."

I had to laugh at that one. It was, again, an unsolicited act of loyalty that my people showed me, and another reminder of why I loved the Marines.

I spent the last two months of active duty keeping busy at my job, but I also had taken Ed Dolan up on his offer to teach me how to golf. Three or four evenings a week that late spring and early summer, we went out to the base golf course and he coached me.

His best advice was for me to buy a stripped down set of clubs (driver, 2 wood, 3, 5, 7, and 9 irons, and a putter, all in a cloth bag) instead of a full leather bag. Golfing on military bases was cheap (I think greens fees on base were $1.50), so we could go out often. Just before Kathy and I were to leave LeJeune, I broke 100 for the first time. A great way to end my active service.

THE ADVENTURE CONTINUES

10.

THE VTU
1972-1973

Back at Mount St. Mary's, where I had been re-accepted as a student, late in the fall semester of 1972 I joined the Marine Corps Reserve Voluntary Training Unit (VTU). This unit met at the Marine Corps Reserve Training Center in Harrisburg, Pennsylvania, on one Wednesday night each month. I had known nothing about the Reserves when I was on active duty, but once at the Mount (in Emmitsburg, MD, just below the Pennsylvania line), thanks to the Marine captain who recruited on campus for the officers program, I found out that while not a high paying proposition, it could be a good deal in the long run. It was, in fact, a no paying proposition with the financial payout coming from accumulated 'Reserve Points,' but paid after I turned age sixty. Every meeting, once a month, counted as four hours of training, each of those sessions representing the equivalent, at least on paper, of one day's pay. While non-paying, service in these units accumulated earned retirement credits.

A reservist had to accumulate sixty retirement credits each year to build a successful retirement program. That was when I started looking at the reserves as an interesting possibility to satisfy my need to be a Marine.

Now how many other people do you know who say, "I think I will work for nothing and then collect a meager pay-check forty years later?" Well, sometimes you find the right reason – the right cause – to which you can make such a commitment. And the rewards, while not monetary, can be pretty terrific.

Col. John Broujous commanded the VTU. He was a very successful lawyer in Carlisle, PA. During the Vietnam War, he took a year off from his law practice, applied for a return to Active Duty, and went to Vietnam to serve a year in combat. He was universally liked and respected.

The VTU (a few years later to be re-designated MTU for Mobilization Training Unit) in Harrisburg had a great bunch of guys who had all served many years in the reserves and were able to give me a lot of guidance about my career. I served with them throughout my time at Mount St. Mary's and stayed connected with them until I went to the Naval War College in 1984.

Each month we would have a meeting at the Harrisburg reserve center. Col. Broujous would preside over structured discussions and training about contemporary Marine Corps issues.

It doesn't sound too exciting, but for each of us, it was a way to keep ourselves in the game.

11.

TEMPORARY ACTIVE DUTY AS A TRAINING ASSISTANT

Summer 1973

As mentioned earlier, during the summer between my junior and senior years at the Mount, I discovered, as I had continued my relationship with the Marine Corps through participation in the Reserves, that I was eligible to apply for a temporary active duty assignment at Camp LeJeune as a Training Assistant. In this capacity, I would return to Active Duty for the summer and become a liaison between the active duty "host" units and the 'visiting" Reserve units coming in for their two-weeks annual training requirement. My main job was to ensure that Reserve Unit logistical support, lodging, and training needs were fully met.

Many Marine Corps Reserve Officers, particularly those who had jobs in education, took advantage of this summer opportunity to pile up some retirement points, and collect a paycheck that was probably a little larger than the one they were receiving as school teachers.

It was a unique summer with lots of things to do professionally, but also I met some interesting people. All of the Training Assistants (TAs), most of us newly promoted Captains, were housed in old one-story, white cinderblock barracks at Camp Geiger, where once, as an enlisted man, I had completed my infantry training and awaited my orders to OCS. Now, in 1973, Camp Geiger had become a little more interesting.

The Marine Corps was experimenting with having the Parris Island Recruits take three weeks at Camp Lejeune (i.e., Camp Geiger) at the end of

Boot Camp to complete their infantry training before going back to Parris Island to graduate. For my TA colleagues and me, this was a treat. Every morning before morning chow, and every evening after we got back from our TA duties on the main base at LeJeune, we got to see recruit troops who were close to graduating from Parris Island doing close-order drill under the supervision of their DIs. For a Marine, that was a great visual pleasure. These were recruits at the end of their training. They had now become adept at accomplishing tricky, and advanced, precision close-order drill. It was a thing to behold seventy-five Marine recruits executing precision drill.

The other advantage of being assigned billeting at Geiger was that the mess hall (dining facility) now served the same high quality meals that were served at Parris Island. (I mentioned in my Parris Island narrative that Boot Camp chow, and that included Marine Corps Recruit Depot San Diego, was the best chow in the Marine Corps).

It was shaping up to be a good summer.

My roommate that summer was a great guy, Capt. Frank Collins. He was just completing Law School at William and Mary College in Virginia. Although from Fall River, MA, Frank and Laurie his wife, and their three-year old daughter were living in Williamsburg, VA, in public housing. Even though his active duty service in the Marine Corps qualified him for the GI Bill that paid his tuition, the Collinses had no other source of income and were living on Food Stamps. I mean, that's just amazing. Here it was 1973, and America had just received the benefit of Frank's active service, yet he and his family had to be "on the dole" as they used to say in Ireland, in order for him to get an advanced education.

The situation with Kathy and me was different. While the GI Bill was paying for my tuition at Mount St. Mary's, I was also working two part-time jobs at the liquor store and as a bartender at a local resort, and Kathy was working full-time as a teacher. Frank did not have the luxury of taking a part-time job thanks to the demands of Law School, and Laurie had to stay home to take care of their little girl. But they were determined to make the experience a successful one.

Both Frank and I recently had been promoted to Captain. As suitemates, we were completely compatible. However, several rooms down from us in our quarters there was an unfortunate pairing of Marine officers that was nothing more than a shotgun marriage in want of a divorce.

Captain Alex Verducci, a high school teacher from Pittsburg, was a really good guy. In Vietnam he had served both as a Motor Transport officer and as a platoon leader in the pacification program called CAP, for Combined Action Platoon. These guys lived in Vietnamese villages, ate Vietnamese chow (which they paid for), and trained the local men in how to conduct village security operations against the Viet Cong. It was a highly successful program, and Alex was very proud of his service.

But to his detriment in our current summer job situation, he was a neat freak. His roommate, another Marine Captain named Casey who was a public school teacher from New York City, was a total slob accustomed to his mother or wife cleaning up after him. This was Felix Unger and Oscar Madison writ large (The Odd Couple was a successful Broadway show, a successful TV series, and a successful movie). The Marine Corps, not knowing (or even caring) about the differences between these two officers, assigned them to share the same quarters. The quarters had a private bedroom for each officer, a small shared living room, and a shared bathroom facility.

The pairing of Alex and Casey was a disaster.

One day Alex, nearly in tears, called Frank and me down to show us the bathroom sink, coated with his roommate's whiskers. You could see he was close to becoming ill. Topping it off for him was the fact that while Alex was totally of Italian ancestry, his roommate was a dyed-in-the-wool, hard-core New York City Irishman who played Irish rebellion music on his cassette player full-blast every night. Alex had complained to Casey, but Casey's response was, "You'll get to love it after a while."

I believe Alex was torn between murder and suicide.

While we sympathized with Alex, there was nothing Frank or I could do. Finally, after he couldn't take it anymore, Alex found a solution. After contacting a local real estate agent, he found a beach house on the Atlantic Ocean at a location called Surf City, just south of LeJeune on the coast. Alex hit a home run with this place. I mean it was right on the ocean. The big bonus was that he couldn't afford it alone, so he offered to share it with Frank and me. Once we saw it, we all called our wives to come down, and Kathy was quick to join us, followed by Frank's wife Laurie and three-year-old daughter Courtney. The remainder of our stay that summer at Camp LeJeune was punctuated by watching the sunrise over the Atlantic Ocean every morning as coffee, eggs, and bacon were being prepared, and coming home in the evening to bathe in the mild waters of the Atlantic.

If I could have afforded it, I would have paid the Marine Corps for the experience.

12.

4TH ENGINEER BATTALION

End of Summer 1973-December 1974

When we returned from LeJeune, I found a paid reserve opportunity for an S-1 position (Administration Officer) in 4th Engineer Battalion Headquarters based in Baltimore. I went down for an interview and was selected for the job. This kind of position was considered a part of the Selected Marine Corps Reserve (SMCR). One got paid two days pay for every one-day of service. That's how the reserves work. So on a typical reserve weekend, I would be paid for four days of service because I had served two days. This sounds very lucrative until you look at the dismal daily rate of pay – even for officers. We did it because we loved 'The Corps.' Down the road, when we turned age 60, this translated into a retirement pay package for us. Not a lot of money, and nowhere near the amount we should have been paid for the crap we ended up going through, but we were happy to do it. We weren't, or at least I wasn't, thinking of retirement pay when we volunteered for these duties.

4th Engineers was a good place to be even if you weren't an engineer. Battalion headquarters consisted of the Commanding Officer (CO), a lieutenant colonel; an Executive Officer (XO), a major; and four "shops:" S-1, Administration; S-2, Intelligence; S-3, Operations; and S-4, Supply and Logistics. Each of the shops, except for S-3, which was led by a Major, were led by Captains. As the S-1, I was responsible for the entire administration of the headquarters unit as well as the satellite engineer companies in Maryland, Virginia, and West Virginia. Battalion HQ maintained all Service Record Books (SRBs) for the enlisted personnel and all Officer Qualification Records (OQRs) for the officers.

When I reported in for my first drill weekend, I was introduced to my enlisted admin chief, Gunnery Sergeant Barbara Deering. She had been the Administrative Chief in 4th Engineers since Jesus was a Private and knew more about administration than everyone. She told me she would handle everything and all I had to do was show up and sign what she put in front of me. The CO recommended that I do just that.

Despite this sage advice, I insisted that the good Gunny teach me about administration or I wouldn't sign one goddamned thing. She taught me a lot that served me well later in my career.

The 4th Engineers had satellite companies in West Virginia, Virginia, and Maryland, each specializing in different aspects of military engineering. One was a construction company, another specialized in road construction, another in the tactical aspects of explosives and mine clearing. They found practical civilian application for their skills by building or restoring municipal parks, building and repairing roads in state and national parks, and renovating buildings and roads owned by non-profit organizations such as nursing homes.

All the other officers in the battalion were friendly and it seemed like I had joined a good family. I got to be good friends with Chief Warrant Officer Frank Wilson who in civilian life was a Maryland State Trooper. He was a great storyteller and regaled us with hilarious anecdotes about his fellow officers and some of the incredibly stupid criminals they dealt with.

I believe I started drilling with the unit on weekends beginning after my return from Training Assistant duties at LeJeune in the fall of 1973 – probably September. Because of the distance from Harrisburg (about 60 miles), I opted to stay overnight in Baltimore on reserve weekends, sleeping right at headquarters. Those of us travelling greater distances were afforded the luxury of sleeping inside the Reserve Center on military fold-out cots. Why waste the tax-payers' money on a decent hotel? (I discovered later that the Army, Air Force, and Navy reserve components budgeted hotel accommodations for civilian lodging expenses for their members on reserve weekends. The Marine Corps had a long tradition of doing business on the cheap because of so many political threats over the years to abolish the service. So fold-out cots it was.)

There was an old Marine Corps adage that said, "We have done so much so long with so little, that next year we will do everything with nothing." This was a hold-over concept from the immediate post-World War II years. President Truman proposed, in 1946, that we might not need a Marine Corps.

As the new strategies for U.S. military force structure were being developed, and ultimately to be finalized in the National Security Act of 1947, the Marines scrambled to garner every bit of legislative support that it could to keep it from being dissolved and absorbed into the Army.

The lobbying effort worked, but until the late 1970s, the Marine Corps was sensitive about asking for additional appropriations for such luxuries as housing for the troops, better equipment, and operations and maintenance funds to keep the wheels turning.

So there was no extra money for motel housing for weekend Marine Corps reservists.

On the Saturday night of drill in September 1974, the 14th to be exact, just before midnight the Duty NCO came in and awakened me to say that my wife was calling. That could only mean one thing. I got on the phone and Kathy told me her water had just broken.

I quickly dressed and raced (in our durable 1971 VW Super Beetle) the 60 miles back to Harrisburg. Fortunately her contractions had not started coming too close together. I got her bag (previously packed for just this event), helped her to the car, and headed 50 miles south down Route 15 to Gettysburg Hospital.

When we moved to Harrisburg in the late spring of 1974, Kathy said she didn't want to switch from the Ob-Gyn doctor she had become comfortable with when we were at Mount St. Mary's in Emmitsburg, so we stayed with Dr. Piper in (back then nearby) Gettysburg.

We arrived at Gettysburg hospital around 2:30 AM and began the long wait. Finally, around 11:00 in the morning, Matthew Damien Power arrived kicking and screaming. I was in the delivery room and it was a wonderfully powerful emotional experience as any father will tell you. I called my mom and dad first with the good news, then I called Kathy's parents, Anne and Ed.

After a few hours the medical folks told me that mother and child were fine and needed their rest and I should go home as it was apparent that I needed mine. I was totally beat having had no more than an hour or two of sleep myself, but I was still amped up.

13.

HOTEL COMPANY, 2ND BATTALION, 25TH MARINES

1974-1979

In December of 1974 an opening for an infantry platoon commander presented itself to me in the newly re-designated H Company, 2nd Battalion, 25th Marine Regiment in Harrisburg, PA. H Company, or Hotel Company as we preferred to call it using the military phonetic alphabet, prior to its new designation, had been what was then known as a Searchlight Battery. This type of unit, which no longer exists in the Marine Corps Table of Organization, had once provided nighttime tactical searchlight support for tank units in combat.

Basically the Searchlight Unit mission was to find a good tactical position on elevations overlooking the battlefield and focus the beams of one-million candlepower searchlights onto designated enemy targets. These searchlights were of the same type that you would see scanning the sky the evening of a major Hollywood movie premier, or the opening of a new car dealership. Each year for annual training, the unit had gone to Marine Corps Base 29 Palms, CA. Upon arrival it would mount its searchlights up on the sides of the Bullion Mountain Range where there was an unobstructed view of the valley where the battle was to take place. The units' job was to await instructions to light up designated targets for the tanks.

In preparation for the main festivities, the troops spent their days playing cards and drinking beer.

At night, they turned on the lights.

In December 1974, the world, as they knew it, just ended.

Advances in technology had rendered these units useless. The new night vision technology, eventually integrated into the M-1 Abrams tank, which was soon to be introduced to both the Army and Marine Corps, made the searchlight batteries obsolete.

So in December of 1974, coincident with the dissolution of the searchlight battery, the unit CO was leaving at the end of his tour, to be replaced by Mustang officer (former enlisted) Major Roger Peterson. The troops in the unit now had to be retrained as infantrymen. A couple of officers who had been homesteading in this cushy unit for years wanted no part of it and transferred out (to God knows where), and so two billets had opened up for platoon commanders. I applied for one of them and Captain Jim Shimp from Williamsport, PA, applied for the other one. We were both chosen.

I should add a note here about the structure of Marine Reserve units. In addition to the core officers and troops and platoons that made up the reserve unit, there was an active duty I & I team in support. The abbreviation 'I & I' stands for Inspector Instructor. Each unit had an Active Duty I & I officer (hereinafter referred to as "The I & I"), usually at this level a Captain, and he had an enlisted team of people that provided administrative, operations planning, and logistics support. Between drills the I & I staff would map out the training plan and arrange for the logistics support we would need during our field training. They also maintained the armory and did a variety of other administrative tasks. More on them later, but for now you should know that we had an ineffective I & I officer, and his staff treated enlisted Marine reservists with open contempt and Marine reserve officers with silent contempt.

Back then throughout the Marine Corps, there was a lot of "us-versus-them" attitude from the active duty people toward their reserve counterparts. And it wasn't just local. When we went to ATD (Annual Training for Duty), our two-week annual training duty, whether it be Camp LeJeune, Camp Pendleton or 29 Palms, CA, the active duty people pretty much treated us with contempt. I saw it first-hand as a liaison officer at LeJeune in 1973, and I saw it every year thereafter as a junior Reserve officer. The active forces, from their top leadership on down, looked upon Marine Reservists as second class Marines.

Over the next several years all that was to change dramatically.

Fortunately I had saved all my instructional manuals from The Basic School, particularly the FMFM-1 Infantry Operations Manual, and in preparation for the new duties I was to assume as an infantry platoon commander, I gave myself an intensive self-tutorial so I could start training my new grunts. It was the most challenging and one of the most interesting, and without a doubt one of the most profoundly rewarding, assignments of my reserve career. Jim Shimp, with whom I developed an immediate bond, and I quickly did an inventory of the people we were to train.

The basic profile of a typical junior Marine reservist (Private through Sergeant) in 1974 indicated that the majority were there because they were able to avoid active service in Vietnam by getting into a Marine Corps Reserve Unit. Very few reserves were called up in that war, so if you could get into any of the Armed Services' reserve programs, you could escape being shot at. All you had to do was get past Boot Camp. We also had a few personnel with extensive active duty service.

Back then, reserve enlistees had a 180 day active duty obligation. They completed Boot Camp and followed that up with a specialty school (Motor Transport, Supply, Engineers, etc.) and were released from Active Duty back to the reserves after the initial six-month obligation. From that point on, all the individual reservist had to do was attend monthly weekend drills, attend the annual two-week summer camp training, and otherwise keep their noses clean and haircuts short and boots shined for the next six years. At the end of six years they were honorably discharged. They also had the option of sticking around by reenlisting every three years for twenty to thirty years and get a modest retirement check from the Marine Corps when they turned sixty years of age. Given the fundamental profile of that generation, a small percentage selected that option.

So when Jim and I arrived, the profile of perhaps 60 percent of the troops in the unit, maybe more, was that of individuals who joined to avoid active service in Vietnam. Most of them were either college graduates or had at least some college. Some were still attending college. They were smart, but they were just marking time and doing their duty. The more junior troops in the unit were newer enlistees who were not under the threat of the draft, or only remotely so. The draft was to end in 1975.

Hotel Company began infantry training in January of 1975. Of our newer junior troops, we had many kids who would have been glad to serve on Active Duty, but were newly married, starting families, and needed to be earning better wages locally. Others simply had better civilian job prospects in general and the reserves served as an augmentation to their wages in civilian jobs. We also had a few former active duty NCOs who gave us some real strength in "middle management."

Within two months, several things became patently obvious to Jim and me. One was that the senior corporals and sergeants, most of whom were within 18 months of discharge, and, unfortunately for the new direction of the unit, were in leadership positions (Corporal Fire Team Leaders, and Sergeant Squad Leaders). They were not trained competently to be leaders. No fault of theirs. As I said, these were a bunch of smart guys, most with some college behind them. There were many fine people within their ranks, but the structure and mission of the old unit had made ardent junior leadership a low priority. We had too many sergeants and corporals in leadership positions who had never been empowered to make decisions at the lowest level (and shame on their leaders). Whenever these NCOs had a problem, they always dumped it in the officers' laps.

To illustrate this let me give you a general example. Haircuts were a huge issue. It was 1975 and the average Marine reservist's vanity said that he had to have long hair to blend in with the civilian landscape. So if he showed up for drill with shaggy locks (or un-shined boots, or a poor shave, or missing gear), the sergeants would write him up and turn the matter over to the company officers. Past practices in the old unit showed that on the typical Sunday afternoon of drill weekend, the CO would spend four hours handing out NJPs (Non-Judicial Punishments, or what we called Office Hours) for hair and uniform infractions.

NJPs were the Uniform Code of Military Justice system's method of dealing with minor infractions without resorting to more severe solutions such as Courts Martial. NJPs were always in the form of reductions in pay, and possibly a rank reduction. These NJPs mostly boiled down to a $15 fine for a bad haircut/shave/boot-shine/dirty rifle, etc. The hearings were held and witnesses were called. It was a waste of valuable time that should have been spent on making the unit a better one by doing some detailed planning for the next month's training evolution.

For kids that were making maybe $45 to $60 per month, and who had enlisted mainly to avoid the draft, sacrificing $10 to $15 per month for minor infractions while they rode out their time meant nothing.

Jim and I put our heads together and in February, 1975, went to our CO, Major Peterson, and proposed that we put the onus of enforcement on the sergeants and corporals. He agreed. Jim and I informed the company that there was a standard to be upheld and that the NCOs (non-commissioned officers, i.e., corporals and sergeants) would be held fully accountable.

This brought about a reduction of administrative nonsense that the CO had to deal with, but there were other underlying factors that had to be addressed. Once again Jim and I went to Major Peterson and said, essentially, "Give us a green light, and you will never again have to deal with that level of administrative trivia. Roger Peterson was simply finishing out his career and would retire within eighteen months, so he green lighted us to run our platoons as we saw fit.

Our plan was this: we went to each of the existing sergeant and corporal billet holders and made them an offer they couldn't refuse. Jim and I each had a cadre of sharp, motivated, up and coming Lance Corporals and Corporals, post draftee era people not threatened by the war in Vietnam, who were motivated to be Marines and hungry for leadership opportunities. We asked each of the existing squad and fire team leaders – those who were finishing their enlistments over the next 18 months -- to step away from their billet responsibilities, yet retaining their ranks and pay, while we turned the platoon leadership responsibilities over to these new kids. To their credit, they all agreed.

Jim and I had easily spotted several terrific young Marines that we wanted to groom. He had a Sergeant Schultz and a Corporal Smith, and I had two corporals, Weikert and Webb, that I wanted to take over squad leader responsibilities. We also had several Lance Corporals who had shown some promise. We got the new squad leaders to understand that they needed to identify others with potential leadership skills who could take over fire team leadership duties.

We told each of the existing leaders that the old ways were finished. We carefully explained that we had identified new leaders who wanted to be squad and fire team leaders, and we told them that we wanted to develop these new candidates.

We were not terribly surprised when the old billet holders said, "No problem. We're getting out anyway and won't stand in the way of your plans."

We told each of our newly selected leaders that we were transferring the responsibility for proper haircuts, boot shines, and all other aspects of Marine Corps grooming and other matters (e.g., rifle cleaning, field gear maintenance) right in their laps. If they wanted to be Marine NCOs, then they had to start acting like them. Good NCOs do not run to officers for all the answers.

All of these new leaders rose to the occasion.

By April we were starting to click on all eight cylinders. NJPs were down to a trickle. The unit was coming around, and each of our weekend drills focusing on Marine Infantry skills at Fort Indiantown Gap, an Army Reserve training center about 22 miles north of Harrisburg, were producing better results. The team was coming together, and fortunately for me, I was assigned a new man, Lance Corporal Bobby Strawser.

In my 25-year career as both officer and enlisted, there are a handful of Marines, both active duty and reserve, whose memory stands out for me. Sgt. Bobby Strawser stands at the top of the list.

He had served in Vietnam as a Grunt (infantryman) and become a corporal. He saw intensive combat in the Ashau Valley where, he told me, conditions were beyond primitive. In defense of some high ground for an extended period, he and his platoon had had no amenities for months. No hot chow. No showers. No clean changes of clothes. The only resupplies were ammunition and field rations, then called C-Rats. After a few weeks in the steamy tropic oven of Vietnam their underwear rotted off. But every day at 1700 (5:00 in the afternoon), for about a half-hour, it would rain like crazy. In anticipation of this meteorological blessing, as the hour approached, each two Marine team manning a fighting hole would strip naked. As soon as it started raining one of the two Marines would jump out of his hole with a bar of soap and scrub himself down while his partner scanned the hostile territory for bad guys. As soon as the first guy was rinsed off, he would jump back in the hole, man his weapon, and his partner would repeat the drill. When the rain ceased after 25 minutes or so, they would air-dry and put their uniforms back on. Every night the enemy attacked.

And you think your job sucks.

(A brief note here: Marines called them fighting holes and the Army called them foxholes. In my experience, Marine leadership always said that foxholes were where small animals hid from predators. In the Corps, Marines dug out places from which they could fight.)

Bobby Strawser had been discharged from Active Duty at the end of his enlistment in 1971. At that time he was a corporal. He married, started a family, and tried to make it as a civilian. After a few years, he realized that the Marine Corps Reserves offered him a chance to pick up an extra $60 per month. Having been out for over three years, the Marine Corps approved his application for reinstatement in the Reserves, but with a reduction of one grade from corporal to lance corporal because of the lengthy break in service. Nevertheless, Bobby jumped at the chance to pick up a few extra bucks each month for his family.

As soon as he joined my platoon I realized that I had won the lottery. He was both a good guy and a treasure trove of information about training infantry troops. After sitting down with him for an intake interview, I realized that I was dealing with a young man of extraordinary talents. He was a goldmine of information. I didn't want him limited by running a four man Fire-Team. I wanted his expertise for the whole platoon so I made him a floating instructor directing him particularly to mentor corporals Weikert and Webb. These were two incredibly promising young NCOs who had exhibited undeniable leadership traits.

I should mention here that I had broken out all my Basic School manuals on infantry operations and had been studying them religiously. I tried to apply all those lessons to my new flock. But there was no substitute for the real thing. Sgt. Strawser had been there and done that. And he flat out knew how to teach it to these green troops.

But, as I had told him, his main task was to train me. He more than rose to the occasion.

Within two months I got him meritoriously promoted back to Corporal, and the following year got him meritoriously promoted to sergeant.

Bobby Strawser was, to my mind, a gunslinger. For those of you reading this with no knowledge of the old American West, I'll try to explain. He was a man of few words, and very soft-spoken, as we used to say. About five foot ten, lean but muscular, Bobby moved quietly, but with a decided grace. He was a man who didn't belong in civilian clothes. (In the civilian world, he was trying to make it as an insurance salesman. I once saw him in a business suit. He carried himself like he owned the building. But that life wasn't for him. To me he was a born warrior.)

Like all gunslingers, he had this casual, confident demeanor that left you with the impression of hidden lightning. He didn't scowl or put on a tough guy persona. With his self-assurance he didn't need to. The real gunslinger never needed to. But those who encountered him knew that there, coiled, and just below the surface, lay buried power. My Boot Camp and Infantry Training friend Jimmy Cooper had that quality. You just looked at guys like that and kept a fair distance. In many cases, when one of these people walked into a room, wiser folk got out. In literature, the only character I could compare Strawser (and Cooper) to was Shane (i.e., the novel 'Shane' by Jack Shaefer). Read it and understand. This type of guy could be your best friend or your most dangerous enemy.

Despite his combat history, Bobby bonded with these young kids who had no warfighting experience. These new troops were the new Marine Corps inductees who were going to replace those who got lucky and joined the unit to avoid service in Vietnam. Strawser had a sincere empathy for them and soon got them to a level of proficiency that was exceptional. They were pumped up and emotionally involved in their training.

It takes an exceptional amount of repetitive learning to become a proficient Marine infantryman. For example, there are a myriad of formations you can use when advancing toward the enemy in combat, and you needed to learn them all, and learn which of those formations would be the best one based on the Mission to be executed, the Enemy you are facing, the Terrain you are traversing, the Time you have available to accomplish the mission, the Time of day, and a myriad of other factors (METTT, yet another Marine Corps acronym). Corporal Strawser taught the troops how to do their own METTT analysis so that they would be ready to take over a squad in combat.

That was one of the things I admired about him. He was focused not only on training the troops in tactics, but also on preparing each of them for the real-life probability that they might, themselves, have to take over other roles once in combat..

A proficient infantryman must coordinate his actions with each of the Marines around him. Get too far ahead, too far behind, or drift laterally in one direction or another and you are, one, not being in a position to cover and protect your brother Marines, and two, putting the lives of your brother Marines in jeopardy. Sgt. Strawser was sensitive to the fact that these kids were raw and untrained. He showed extraordinary patience while bringing them to a level of proficiency that would not only rival but exceed the capabilities of their active duty counterparts, as was to be proven later.

Soon we began clicking on all eight cylinders. And it only got better.

We did all of our monthly field training at Fort Indiantown Gap, PA. The 'Gap' was primarily a Pennsylvania National Guard and Army Reserve training base located twenty-two miles north of Harrisburg on Interstate Route 81. It was nestled on an eastern slope of the Appalachian Mountains, not far from the Appalachian Trail. The Army Reserve and National Guard were very supportive of our need to conduct realistic training, and while initially the support took mainly the form of allowing us access to training areas, in another year or so, we were given a much higher level of support.

Each month our unit trained in squad (13-man) and platoon (42-man) tactics, and occasionally as a company (ideally three infantry platoons of forty-two men each, but in our reserve unit it was two platoons of 42 men each). That would improve in coming years.

One of the 'motivational' squad level training events at Indiantown Gap involved the "slide for life." This was a challenging, confidence building event that required our Marines to climb a fifty-foot tower, and hook onto a cable that would deposit them at ground level some 225 feet away. (Today every resort in the country has "Zip Lines" where you can get your adrenalin high; sensible travelers avoid them.) For most of the troops, while it was a new idea, it was no big deal. But when it came to Sgt. Weikert, one of my chosen new leaders, and a young man with loads of leadership potential, it became a challenge. He was terrified of heights.

With that in mind, I made sure that he and I were the first two in our platoon to do the slide. I climbed up first and he followed. On that platform I talked with him and told him that if he was to become a leader of Marines, he must overcome his own fears in order to set the example. He watched as the instructor hooked me up, and I looked at Sgt. Weikert and said something like, "See you at the bottom," and pushed off. For me, it was a fun thing to do, and when I was assisted off the wire at the bottom, I looked back up to see that Sgt. Weikert had just stepped off the platform. I was proud that he had the courage to launch himself into that which he feared most. It was a terrific step on his journey to becoming a top-notch squad leader.

In August, for the first-time doing infantry training, Hotel Company went to Camp LeJeune with Headquarters, 2nd Battalion, 25th Marines, for two weeks of training and tactical exercises.

One of the great drawbacks of reserve training is that you only get to train with your entire parent organization (both battalion and regimental staffs) once every three years – occasionally more frequently, but that was the structure of the training regimen of Marine Corps Reserve at that time. During the first week of our training in 1975, Hotel Company and Battalion Headquarters for 2nd Battalion, 25th Marine Regiment were at the Infantry Training Regiment at Camp Geiger for advanced training in infantry tactics. This was the same place where I did my initial infantry training as a Pfc.

Our battalion consisted, as did all Marine Corps battalions, of three rifle companies. There were two other infantry companies in the battalion, and they would, sequentially, get their opportunity at training with battalion headquarters in coming years. The hope was that enough troops and NCOs would re-enlist so that when it became our turn to train with Battalion again four years out, we would retain some corporate memory of our previous experience. That was one of the limitations of reserve training, not just for Marines, but for all the services. Our entire battalion should be learning how to operate together, as we assumed we would be mobilized together. However to effectively train a full battalion of three rifle companies with all the concomitant battalion headquarters coordination of close air support, artillery, Naval gunfire, tank, engineering, and logistics support would take six to eight weeks if not longer.

Early that first week of Annual Training Duty (ATD) in 1975, the 25th Marine Regiment operations officer, Major Flannigan, assigned Hotel Company the responsibility of providing a raid platoon for the big exercise that would happen the following week. Our company commander, Major Peterson, was reluctant to choose one of us platoon commanders over the other, so Jim Shimp and I each selected from our platoons thirty-man teams to train for the raid. We trained competitively that first week to see which of our teams would get the honor of doing the raid.

Both Jim Shimp's platoon and mine trained hard for the honor of being selected. The scenario, (a night-time raid on a SAM (Surface to Air Missile) site (simulated) had been given to us and we trained in daylight on the site where the actual raid would take place for three days. Jim's team and mine alternately took turns in rehearsal.

At night, Jim and I relentlessly went over the details of the raid. To prepare, Jim and I went to the CO of the Training Command and, per our request, he provided us with the official Marine Corps instructions for conducting raid operations. Late into each night Jim and I, though competing for the honor of leading the raid, worked together to produce a detailed and comprehensive Raid Order. Whichever of us was chosen, the order was to be the result of our best collective thinking.

A Raid Order (it seemed to me), more than other types of operational orders in the Marine Corps, required an attention to detail that boggles the mind. Not only was the objective clearly defined in minute detail, but the execution was broken down very specifically as to what weapon(s), uniform, equipment, and other related gear every single man would carry.

The Raid Order specified everything every man was to carry down to the last band-aid. There is a copy of this order in the cardboard box with all my Marine Corps records. It's worth saving.

Jim and I both knew that only one of us would be designated as the raid commander, but we retained a healthy sense of both competition and cooperation. We both wanted Hotel Company to look good. We knew that the active duty folks would be watching.

Finally, on Wednesday evening, after three days of intensive rehearsal, one of us was to be chosen. The CO, Major Roger Peterson, wanted no part of making the choice (basically he bailed out on his leadership responsibilities and pulled a Pontius Pilate act of virtually washing his hands), so he designated the XO, Captain Vic Alt (more on him later) to make the choice. Vic was a close friend to both Jim and me. I had worked with Vic at Pennsylvania Blue Shield, and Vic knew Jim from active duty. As a matter of fact, Jim and I lobbied on Vic's behalf to have him join Hotel Company. But Vic, like Roger, was reluctant to make a decision. So he finally wussed out and defaulted to a coin toss. Both of our platoons had done so well in rehearsing for the operation that either of our units would have excelled at the task. But only one officer and thirty Marines would be selected.

I won the toss, and out of a sense of fairness, and after discussing it with Jim, brought half of his platoon in to join mine. A lot of the guys from both my and Jim's original thirty, while having executed their training assignment well, really didn't care whether they did the job or not. They were among those I have previously described as just waiting to finish their time. They were good people, but I wanted the most motivated to go on this mission. So I asked Jim to help me combine the best of his unit with the best of mine. Excluded were the soon to be discharged sergeants and corporals. I rehearsed the raid scenario with our new hybrid raid platoon many more times, and then it was early Friday evening and training was done for the first week.

A funny thing happened on that Friday evening. Hotel Company was encamped in a "hard back" tent area, that is, each tent sitting on a concrete slab held a dozen or so cots. Given the August heat, the sides of the tents had been lifted for air flow and comfort. Our raid unit had just come back from our final rehearsal and the troops were incredibly pumped.

After a field shower and chow, we settled in for the evening. The officers and senior enlisted held a confab to discuss lessons learned from the first week, as well as plans for the next week. By the time we wrapped up around 2130, it was getting dark. Suddenly we became aware of a low chanting in the background and looked across the troop billeting area to see candles burning in one of the squad tents.

Curious, we walked over to see what was going on. It seems that the NCOs that we had relieved of duty (the short-timers who were getting out in the next several months) were, while wrapped in sheets, holding a combination wake and séance. When Major Peterson asked what they were doing, one of them said, "Captains Power and Shimp have ruined everything. The old organization was free-wheeling, no sweat, and not much work. Now they have everyone pumped up and running all over the boondocks and getting excited about everything. We were just mourning the loss of what we knew."

We officers just shook our heads. Major Peterson wisely said, "Carry on."

I believe it was Sunday afternoon that we boarded CH-46 helicopters (these birds have two big rotors, one forward above the cockpit, and one elevated above the tail) for our trip out to sea to board the USS Guadalcanal, a Korean War era carrier that had been converted to helicopter operations. The Guadalcanal was one of a fleet of four ships that were to deploy the entire battalion, along with the regimental staff, for the three-day exercise at LeJeune. It was my first time going aboard ship by helicopter and it was a little disconcerting. Several miles out to sea in our 46, I could see through one of the side ports that we were approaching the ship. It looked as tiny as a postage stamp and I couldn't conceive of our landing on it. Finally we approached the port (left) side from an oblique angle, and the pilot, timing the up and down pitch of the ship, lifted the helo from a side angle off the port quarter up onto the deck for a soft landing.

Once aboard, we conducted final briefings and inspections to make sure each Marine knew exactly what to do and had exactly what he needed to accomplish the mission once in the combat zone. After evening chow aboard the ship, I sent everyone to bed early as we had a 0130 wake up time. The Guadalcanal, which was quite an old ship at the time, had no air conditioning below decks for the Marines, though the Navy officer and Navy enlisted quarters had plenty of cool air. At that time, and for the next two years, we realized that this was Navy unofficial policy. Basically it meant, "Don't spend any extra money for the Marines as they are only on board temporarily." It was in keeping at the time with the Cold War strategy of focusing all efforts – and Congressional appropriations -- on the "blue water" (i.e., deep water) based Navy that was centered on fleet protection for the aircraft carriers that handled the national policy of prepping for war with the Soviet fleet.

Amphibious ships that Marines used to deploy troops ashore were considered "brown water Navy" and given second-class status within their own service. What the great strategists and force planning theologians had failed to grasp was the fact that if you couldn't put the little green man on the little green hill, you were just jerking off.

The air conditioning was broken, and I found out that it had been broken for quite some time before they knew that they would be embarking our Marines (which was bull shit – it was an amphibious goddamned ship and its freaking job was to embark and deploy Marines), and for the troops it was a hell-like sauna. Imagine the hottest, most humid summer night you have ever spent without even a low-level fan blowing. Now imagine that same summer night inside a garbage can. That doesn't come even close to describing the inferno below decks of an un-air conditioned amphibious ship full of infantry Marines.

These ships were, at least above the water level, no more than a three-quarter inch thick piece of metal. In the North Carolina waters off of Camp LeJeune, they absorbed heat. My guys sweltered below decks while the Navy crew slept in coolness.

Any reason I should be pissed?

As often as I could, I was down in the troop compartments talking to my Marines and prepping them for the operation.

I did have to spend quite a bit of time in the officers' ward room (air conditioned, of course) fine tuning the final plan and talking with the helo pilots making sure everything was coordinated with the two flight crews that were to bring us into the LZ (Landing Zone) in the pre-dawn hours. My fellow officers from the helicopter ranks and I also conducted a detailed review of the communications links with the Recon team on the ground that would guide us in.

I was a little put out that I never had a chance to coordinate directly with the Recon team. What had happened was the Regimental staff dealt directly with Recon as the team was considered a regimental asset. Regiment had given us reassurances that Recon was fully coordinated with our plan.

This became a "lesson learned" moment for me as you will soon see.

I slept in my quarters for about an hour, then went down to the troop compartment at 0130 to get them up. It was beyond a fucking sauna. It was on its way to becoming Dante's "New and Improved" Inferno. But the troops were already up, cocked, locked and ready to go, as they say. The troops couldn't sleep anyway, so I moved them up to mess deck for chow (traditional steak and eggs – the US Navy's traditional gift to those who are about to die for them while they drive around the ocean and drink coffee). Soon we moved into the passageway leading to the flight deck.

I, as well as our Gunnery Sergeant, conducted the umpteenth inspection of every man and the gear he was to carry on this mission.

My raid platoon was to embark aboard two CH-46 (dual large rotor) helicopters in the pre-dawn at 0300 for our raid objective. The goal was to go into the "hot" zone guided by an advanced Recon team, land, and take out a (simulated) S.A.M. (Surface to Air Missile) sight in LZ (Landing Zone) Albatross, near the beach at Camp Lejeune.

We had prepped for this moment intensively. The troops, in spite of their discomfort below decks, were pumped up and ready to go. Every bit of their exposed skin was camouflaged, and all these fired up Marines wanted to do was to put into action all that we had trained for so intensively the previous week. They were running on adrenaline. Along with the senior enlisted on the mission (a Gunnery Sergeant and two Staff Sergeants), we conducted a final detailed inspection before boarding. Once aboard the two CH-46s, we lifted off. In pre-dawn darkness we flew in low over the ocean (50 feet altitude) and popped up just above the tree line to swing around and land at the designated site.

As we swung in over the LZ the pilot of our bird motioned me to come to the cockpit. There was a problem. Our primary means of communicating with the Recon Team on the ground was what is referred to as a Black Box. The Black Box was a homing device that our helicopters could lock in on and be guided to the LZ safely. The pilot told me that there was no signal from the Black Box. Our secondary means of communication was the radio, but we could not bring up the Recon Team on the assigned frequency. The third method of communications was to be strobe lights that the pilots could guide in on. The pilot pointed down to the LZ and all I could see was strobe lights all over the place. It looked like a massive Christmas display. Being a training exercise, the powers that be had left the defense of the site to an engineer unit. Every individual Engineer carries a strobe light. It is part of their Table of Equipment.

Over the noise of the engines, I made a slashing motion with my hand near my neck that told the pilot to abort the mission and he looked at me with relief. Disappointed, we returned to the Guadalcanal. The regimental Operations Officer directed us to fly in administratively a few hours later and join up with the rest of Hotel Company in one of the landing zones near the beach. The rest of the company had landed tactically by landing craft that morning at first light.

As soon as we landed, I took a jeep to the Regimental field headquarters. Major Flannigan, the S-3 (Regimental Operations Officer) debriefed me. As it turned out, the Recon Team had lost the black box overboard from their rubber boats as they were slipping into the landing zone. Secondly, either the battalion or regimental communications officer had given the Recon Team the wrong radio frequency, which was our secondary back-up communications channel. Third, the engineers defending the LZ not only had strobe lights, but they had constructed (simulated) pole mines throughout the zone. Pole mines are mines that are mounted on six-foot stakes and designed to trigger an explosion when a helicopter attempts to touch down. Even though these were simulated, this is the information that the umpires had. Had we attempted to land the helicopters, the exercise umpires would have declared us all dead.

I told the S-3 that my guys had been unbelievably pumped and were sorely disappointed. He had overseen many of our raid training evolutions and was sympathetic. We hoped he could plug the raid back into the training exercise sometime later. He said that he would do what he could. I went back to the LZ to rejoin Hotel Company, and that's when we got a hilarious recounting of the amphibious landing from the perspective of one of the Training Assistants assigned to our company. (You might remember that between junior and senior year at the Mount I did summer duty at LeJeune as a TA.)

It seems that the landing was to be a major media event, and bleachers had been set up on the beach for families, local Marine brass, Congressional VIPs, and television cameras to review the proceedings. There were hundreds of people in the stands. – wives, kids, Congressmen, Generals, etc.

Well, here came the amphibious landing craft in a proper line headed for the beach. All, that is, except one. The main body of World War II vintage

AMTRAKS (Amphibious Tractors – they were ship-to-shore troop vehicles that transported Marines from their ships to combat zones, engaged their tracked tread upon reaching land) landed, pulled up onto the beach, and disembarked their load of combat Marines, who went charging into the woods.

All well and good, and good news and video copy. However, one straggling AMTRAK that had fallen behind the line was belching black smoke and trying to chug its way to the beach. Let me add that once Marines are embarked on these craft from the ship, the cover of the troop compartment is closed until the craft hits land.

So here comes this relic of World War II (at that time Marine Reserves always got the hand-me-downs from the regular forces, so we were using equipment from World War II – this was 1975 and the war had been over for 30 years, by God!).

This ancient AMTRAK was belching black smoke and lurching its way through the surf to give its final gasp and come to eternal rest on the sands of Onslow Beach, Camp LeJeune. The rear hatches popped open and along with a billowing cloud of black smoke its cargo of Marines blasted out of its interior like they had escaped a level of hell. They staggered out coughing and retching, dropping to their knees and trying to regain their senses.

One Marine was so blinded by the smoke that he unknowingly staggered over in front of the VIP bleachers coughing like mad. He slammed his rifle down in anger, dropped to his knees and puked his guts out. Wiping his sleeve on his mouth, he stood up, undid his fly, and began urinating right in front of the wives, kids, generals, cameramen, and Congressmen. Women, their mouths agape, tried to avert their children's eyes; General officers gritted their teeth in embarrassment; Congressmen looked on in wonder; and junior officers tried to stifle their laughter. Finally looking up, the Marine calmly closed his fly, picked up his rifle, saluted, and disappeared into the woods.

You can't make things like this up.

And of such things legends are born.

The rest of the day Hotel Company remained in the defense while the two other rifle companies attacked various objectives.

The next day, fortunately, the Regimental S-3 had readjusted the training plan so that we could, at last, conduct the raid.

At 2200 (10:00 o'clock) Wednesday night we boarded the helos once again. This time everything was flawless. The helicopters hit the LZ (the pole mines had been "removed") and we were off the aircraft so quickly that the ground observers, active and reserve officers, including Major Flannigan, the S-3, never saw us disembark. The helicopters quickly ascended and disappeared in the dark.

I had broken the unit into three elements, an assault team, a command team, and a security team. As soon as we were off the aircraft, the security team, under the command of Sergeant Weikert, deployed about three hundred meters to the right and assumed a prevention defense position at the entrance of the only road leading into the zone. I led the command team -- a gunnery sergeant, a radio operator, the medical corpsman, and me – and deployed behind the center of the assault team directed by Sergeant Smith. The two assault teams were each led by Sergeant Webb from my platoon and Sergeant Smith from Jim's platoon.

In less than five minutes, the assault team had laid down a base of fire, deployed sapper teams to blow up the S.A.M. site, and, following the final explosions, began the hopscotch withdrawal plan we had rehearsed so extensively. As we withdrew, over the radio I gave the command to the security team and Sgt. Weikert began pulling his team back while providing covering fire for the main body. We pulled back to an extraction point 300 meters to our rear. We had been in the hot zone less than 10 minutes.

On the run I called to the helicopters to execute the extraction plan. Reaching the extraction point, I deployed the troops in a 360° perimeter. Sgt. Weikert's people came through our lines and shored up the defense. Right on time the birds dropped into the center of the circle and extracted our people promptly. We were flown back to New River Air Station, adjacent to the Infantry Training Regiment at Camp Geiger.

Deplaning, we were brought up to one of the training rooms for a debriefing. I never saw anything like it. The troops were off the wall on an adrenalin high knowing that we had so perfectly executed the raid plan. Our adrenalin was coursing through our bones and we couldn't stop our back slapping, shaking hands and hugging. When things finally settled down, we had a good debriefing from each element. Each Marine did his job exactly as rehearsed. The new NCOs that Jim Shimp and I had groomed had turned the company into a motivated fighting team that could rival any active duty Marine unit.

The next day I debriefed Major Flannigan on our perspective of the raid, and he filled me in on what it looked like from his end. He and several observers were in the zone to watch the entire event unfold. He was incredibly effusive in his praise. I was to find out more of his praise a month or so later after my fitness report for the two weeks training was submitted to regiment. This report was written as a decently positive evaluation, but with a characteristic lack of enthusiasm by the Hotel Company XO, Vic Alt. Vic was one of those people who was afraid to deal in superlatives fearing that his superiors would think he was overstating the situation. Major Flannigan, however, back at regimental headquarters, added a statement to my fitness report praising, in his words, my "incredible leadership," a leadership he had never before seen in his career. This from a Vietnam combat veteran.

As a post-script to this two week adventure, while on final liberty, before we boarded our Sunday transportation back to Harrisburg, PA, on Saturday night, the raid team (minus me) went out partying in Jacksonville, NC, the town adjacent to Camp LeJeune (and, back then, a hell-hole representing the worst our nation could produce in military towns), and partied hearty. (Several years later I had a conversation with the wife of an old Navy friend of my in-laws. Her husband had left the Navy and became a senior executive with one of our major oil companies. The woman said that of all the God-forsaken places her husband's company had sent them, Jacksonville, NC, was the worst.)

By the time my guys got kicked out of all the bars in J-ville, several had gotten Marine Corps tattoos, and the corpsman, who was a graduate student music major at Penn State, got a small music note tattooed on his shoulder.

Before we left Camp LeJeune, Corporal Strawser offered me the greatest compliment I have ever received in my twenty-five plus years of service in the Marine Corps. He said, "Captain, if we had more officers like you, there would have been fewer fragging incidents in 'Nam.'" He was referring to the unfortunate practice, though rarely conducted, of rolling a 'hot' grenade under the bunk of a sleeping incompetent lieutenant.

I was absolutely stunned by his assessment of my leadership, and I hold that praise of my Marine Corps leadership talent closest to my heart.

Despite the horrible things that happened at OCS, despite the challenges I faced working for Lt. Col. Ray in Iwakuni, and despite other horrible things that would happen later in my career, it will always be Sgt. Strawser's analysis of my leadership skills, as well as an act of respect promulgated by Sgt. Gary Weikert, that I value above all.

For all of us, returning to our civilian jobs the following Monday offered nothing that could compare with what we had accomplished.

* * *

In early October of that year (1975), I got an eye-opening introduction to the concept of loyalty. I had been working at Pennsylvania Blue Shield for about a year-and-a-half, and I was due for some vacation. I put in for a week in the beginning of October. Matthew was only a year old. We planned a trip that included an overnight at the Baseball Hall of Fame in Cooperstown, then a trip to spend a few days with Don Turlick in Mashpee on Cape Cod. Our return trip included a two-night stay with my sister Elaine at her apartment in Danbury, CT, where we would have a special celebration of Matthew's first birthday.

It was a wonderful vacation, but as we were leaving Cape Cod and heading to Danbury we heard on the car radio of a major hurricane, Eloise, that was headed up the Atlantic coast.

Harrisburg had been flooded three years earlier by Hurricane Agnes (two years before we moved in), and in our living room, about two feet from the floor, you could see the water line that the flood left on the gray fireplace stones. We were concerned. Like all the houses in our neighborhood, the one we had purchased was dry, had brand new electrical wiring and new flooring on the first floor. Our house, like the rest of our block, had been thoroughly dried out with de-humidifiers, fans, and portable heaters for six months. Once dry, the houses had been meticulously restored to pre-flood conditions.

Our vacation visit with Don on Cape Cod had been great. The early October weather on the Cape was still pleasant thanks to the ever-present influence of the Gulf Stream. After a couple of days with Don, we headed for Danbury. When we got to Elaine's, we kept an eye on the weather reports. She and Kathy put together a terrific dinner, and Elaine had also baked a cake in the shape of a teddy bear for Matthew's first birthday. He was thrilled and couldn't wait to cover himself in cake crumbs and icing.

The next day the rains came. A small creek that ran near Elaine's condo had overflowed its banks and was climbing up her front steps. We made a call to Harrisburg to our neighbors, the Leitners, who lived across the street from us, to see if they could get into our house and move at least our valuables (silverware, stereo, etc.) to the second floor. There was no answer at their house. Now we were worried. I remembered that Kermit Leitner's father was a physician, so I got telephone Information to track him down. We finally reached Dr. Leitner, who said that Kermit and Connie had fled Harrisburg to stay with them in Camp Hill across the Susquehanna River. Basically, the residents of Harrisburg had, once again, been encouraged to evacuate.

Dr. Leitner put Kermit on the phone, and Kermit promised that he and Connie would do what they could.

About a half-hour later I got a call from Connie. She said that they went to our house and were trying to find a way in when suddenly a window was thrown open and a Marine stuck his head out demanding to know what the hell they were doing trying to get in the Captain's house. Soon, after explanations that assured my troops that Connie and Kermit did not fit the profile of your average looter (she was a trained concert pianist and he was a high school administrator for Christ's sake), they were permitted in. It turned out that Sgt. Gary Weikert had contacted our platoon to meet him at our Reserve headquarters, just up the street from our house, to assist in moving weapons racks to the second floor of that building. When they were done he said, "I know the Captain is on vacation and he lives just four blocks away. Let's go check out his house."

Gary was about twenty-four years-old and had married a very nice woman who was about forty. He therefore had a stepson who was only a few years younger than him. Fortunately, they got along very well. His stepson, who was a Lance Corporal in the unit, got into our house through a pantry window that we didn't even know was unlocked. Once in, he let the rest of the platoon in. By the time Connie and Kermit got there every bit of furniture, the stereo, all of our silver and everything except the wall-to-wall carpeting and fireplace had been moved to the second floor.

Over the next several days, Sgt. Weikert and his stepson checked the house out every four hours after that until the floodwaters subsided. (We never did get flooded, but the river nearly crested its banks just one block from our house.) Before we had returned to Harrisburg a couple of days later, Weikert and the platoon had returned all of our belongings to the first floor.

I was bowled over by that kind of loyalty, and tried my best to continue to earn the troops' respect for as long as I was in the Marine Corps.

* * *

One of the more fun things we instituted in either the fall of 1975 or 1976 was a Marine Corps Ball for the troops. A few of us planned the whole thing out, drew up invitations, arranged with Headquarters Marine Corps for a guest of honor, and invited all area Marines and those who had formerly served in the Corps to the event. It was one of those uniquely military events where the price of tickets was scaled to the rank you held. The most senior officers paid the most and the most junior Marines paid the least.

For a local event, I always thought it was elegant. Those who had Dress Blue uniforms wore them, and those who didn't wore their dress greens. All the wives and girlfriends wore their best formal attire. As prescribed by Marine Corps Custom, there were rituals, speeches, a wonderful dinner, and dancing. We did that every year until I left Hotel Company in 1979. I think the tradition died out then.

* * *

Our annual training in 1976 was the toughest, and most challenging physically, that I ever experienced. The Marine Corps had decided that its reserve component should undergo the same level of training that the active duty infantry counterpart endured. That meant that units must start into a cycle of training that included cold weather mountain training, desert training, and jungle warfare training.

So in May 1976, we were to embark to Jungle Warfare School, Panama. Our 2nd Battalion Headquarters unit which was based in New York City was to fly down in USAF C-141 Stratolifters, at that time the second largest troop transport plane in the US arsenal. We would meet them at Harrisburg International Airport, stow our gear on board, and then head for Panama.

Before that logistics enterprise was to happen, we needed to actually have all of our Marines present. Enter our company First Sergeant, Glenn Laughman.

First Sergeant Laughman was 1) the most well-organized person I have ever met; 2) the most competent senior Non-Commissioned officer I have ever met; and 3) the funniest man I have ever met.

Joining me on our trip to Jungle Warfare School, Panama, were Captain Vic Alt, our new company commander, Captain Jim Shimp, erstwhile leader of the Second Platoon, and newly joined First Lieutenant Jack Kotanchik, recently departed from Active Duty and now our new Weapons Platoon Commander. Jack was new, shy, and somewhat intimidated by our more casual approach to each other. Sometimes it took a while to shake off the rigors of Active Duty.

As we assembled at the Marine Corps Reserve Center in Harrisburg early that Saturday morning, knowing we were to be at the airport by one PM to meet out transport planes coming from New York, roll call revealed that we were missing one Lance Corporal who happened to live forty-miles away in York, Pennsylvania. As the rest of our pre-embarkation efforts were moving along smoothly, at about 0930 1st Sergeant Laughman said he would go get the missing Lance Corporal.

Three hours later, as we assembled on the tarmac at Harrisburg International Airport, 1st Sgt. Laughman's appeared with a scraggly looking human specimen, more-or-less in proper Marine utility uniform, dragging a seabag behind him. Apparently our missing person had decided to go out the Friday evening before our departure and returned home at around 0300 totally trashed.

When 1st Sgt. Laughman arrived at said Lance Corporal's home late that Saturday morning and rang the doorbell, the miscreant's father simply opened the door and pointed 1st Sgt. Laughman up the stairs. There, in a fabulous mess of twisted bedding and a floor full of dirty laundry lay our culprit.

Our good 1st Sergeant grabbed the first ear he could find, squeezed it enough to get the attention of this near-comatose moron, and dragged him out of bed. The miserable wretch never knew what hit him. Before he could put a coherent sentence together, Laughman made it perfectly clear what was required. Our miserable wretch then scrambled to assemble whatever military gear he would need in Panama. Once he was more-or-less organized, Laughman dragged his sorry ass down the stairs, along with his hastily packed seabag. The little wretch's father simply stood and held the front door open while Laughman dragged this idiot's silly carcass to his car and threw him in.

Happy to see our beloved 1st Sergeant, we embarked from Harrisburg International airport in Air Force C-141 transport planes and headed for the Panama Canal Zone and the Jungle Warfare School located there at Fort Sherman. The C-141 was a monster of an aircraft capable of lifting several hundred troops and their gear and transporting them several thousand miles. While not the biggest transport plane in the Air Force inventory at the time – the C5A was the biggest back then – the C-141 was bigger (and, I might add, louder) than anything any of us had ever flown on. It was certainly larger than any civilian airliner in existence, and that was going some. The 747 commercial jet was a huge airplane, but the 141 was much bigger, and when you saw the C-5A, you wondered that it ever got off the ground.

This monster of a plane had its wings over the top of the body as opposed to commercial airliners that had the wings underneath the plane. This was a common characteristic of all military transport planes that distinguished them from civilian aircraft. Hanging below the huge wings were four humongous jet engines. Inside the aircraft there was not one shred of insulation. The walls of the plane had wires and hydraulic lines totally exposed. And with no insulation, the noise inside was deafening.

We had been issued foam rubber earplugs and we dutifully stuffed them into our ears.

With a stop for refueling at Charleston Air Force Base in South Carolina, we commenced the remaining eight-hour leg of our flight to Panama.

We arrived at Fort Sherman, Panama, late on Saturday evening, and went through extensive briefings on Sunday to prepare us for two extremely intensive weeks of physically draining training. We were the first Marine Corps Reserve unit to undergo Jungle Warfare training in Panama. The main problem we faced was that our annual training was limited to two weeks, yet the Jungle Warfare curriculum demanded three weeks for successful completion.

No problemo.

The Marine Corps simply told the US Army to cram the three-week training program into two weeks. The way that was accomplished was by eliminating a great number of hours of sleep.

In preparation for this experience, following our briefing on Sunday morning, Jim Shimp and I went on a three-mile run. We chatted about the challenges before us, and we had a good workout.

So Monday morning bright and early we broke into squad sized units and began the daylight land navigation portion of the training. Each land-nav unit was composed of 12 to 14 Marines and their officers. The land-nav course had three legs of roughly 1200 to 1500 meters each, and each one of these 'legs' was navigated through triple canopy jungle. All of us were armed with machetes and we each (officers included) took a ten to fifteen-minute rotation on point as we hacked our way through the jungle. Any more than ten to fifteen minutes on point hacking away at the jungle would mean a medical evacuation. Each squad leader was given a compass heading for each leg and it was up to him to guide the point man in the right direction.

As it turned out, Jim Shimp became a heat casualty and ended up crawling out about 300 yards through the jungle where he was assured he would find a road and someone would come along that road and pick him up. Well that's exactly what Jim did. And when he got back to the barracks the Navy Corpsman put him on an IV to restore his, well, everything. He was restricted to 24 hours of light duty.

For the rest of us, the heat and humidity were unbelievable. As Kipling once said in Gunga Din, "... the 'eat would make yer bloomin' eyebrows crawl." Before we completed our three legs at around 4:00 in the afternoon (we had started around 6:30 AM), two of our squad members had to be medically evacuated because of heat exhaustion. That was about the score for each of the nine squads in our company.

When we came out of the jungle late that afternoon, soaked in sweat and filthy, we arrived back at the company area, which was right along the Atlantic side shore of the Gulf of Panama. We dropped our weapons and related military gear on the beach, and in our exhaustion, marched straight into the refreshing salt-water and sank down up to our necks. The two guys who had been medivac-ed that morning were fully recovered (both having been the recipients of IVs), and they marched into the water carrying trays of cold beers to us as we wallowed in the shallow waters of the Gulf of Panama. (That's another unique Marine Corps image in my brain that has no civilian counterpart.) Finally we went in to shower, change into dry uniforms, eat a quick chow, and get ready for the night land navigation portion of the training.

As the remaining officers assembled in the barracks in our now clean, dry utility uniforms, we stopped by Jim Shimp's rack to give him a word of encouragement. He was sorry he couldn't join us on the night leg of the land-nav problem, and we assured him that it was okay, no problem, hang in there, you'll be back out there tomorrow. Then shy, recently released from Active Duty First Lieutenant Jack Kotanchick leaned down to Jim on his bed and said, clearly so we all heard it, "Wimp."

We all roared with laughter at that one..

Our first leg of the night navigation began at twilight. We were to hack our way through 1500 meters of seven-foot high elephant grass (razor-blade sharp tropical grasses that grew to these heights while remaining about one-inch wide with sharp edges). In this first leg, the "grass", if you could call it that, was tough to cut, even with a machete. We finally figured out that if the lead man simply fell forward and bent it down to the ground, we could make faster progress than by hacking at it.

Once we accomplished that simple task, we proceeded to the second leg. It now was dark. I mean really dark. Take away every bit of light you can imagine, and you had Panama jungle dark. There wasn't a smidgen of light coming through the dense triple-canopy jungle. We navigated through this pitch darkness on a different compass heading for another 1500 meters to attack an objective defended by army soldiers.

That leg was a killer.

Initially we could still see the stars. All too soon we got into the jungle where it was beyond pitch black. All we could do was hang onto the belt suspender straps of the man in front us and keep moving following whoever was in the lead rotation. We had given each man a luminous strip of tape to put on the back of his cover (hat), and it was that upon which we guided our unit through the jungle. Whoever was named point, and that role changed every ten-minutes, was dialed in to a compass heading that, like wristwatches of the time, had irradiated illumination dials. I have never experienced such a light deprivation event. It did little to assuage my deep-seated claustrophobia. It took everything I had over those final seven hours to keep from freaking out. Total darkness terrified me.

At around 1:00 in the morning, after seven hours of hacking our way through triple canopy jungle, we finally broke out onto the dirt road that would lead us to our objective. Surprisingly, we came out of the jungle exactly where we should have. Exhausted, we counted heads – miraculously we discovered we had lost no one. We got into tactical formation and attacked the objective. Our target was defended by US Army Rangers who graciously let us win. They wanted to go back and sleep in their barracks just as much we did. May all wars end this way.

Around 2:00 AM, following our successful vanquishing of the enemy stronghold, we formed up on a remote dirt road to await the trucks that would bring us back to our base at Fort Sherman. As we waited there in military formation, suddenly the temperature dropped about five degrees and the skies opened up. It started raining. It wasn't just rain as you and I know it. It was Second Coming rain. It was Noah Biblical rain. These were raindrops of a stunning size -- rain like I've never seen before. The drops were the size of golf balls and in seconds we were all soaked. I turned to First Lieutenant Jack Kotanchek and said, "This is perfect."

He looked at me, soaked to the bone, rain dripping from every surface of his body, and said, "I'm only doing this for the money."

Again one of the great moments in my Marine Corps story where I got laughing hysterically. We weren't ever going to be repaid for the value of what we had endured.

I do remember that we ended up with about two hours sleep that night, and that was to be about the most we would get in any night for the next two weeks.

Our days were filled with training classes out in the field. In an outdoor 'classroom' where the real stuff was happening, one with a vivid imagination might argue that classes were almost academic in nature. We were seated on benches and our instructors were in a stage area. The difference between our classrooms and your basic college classrooms was that we sat in 99 percent jungle induced humidity, emphasized periodically by invasive insects whose mission was to devour us. Our Army instructors educated us about the flora and fauna of the jungle. In those sessions we learned what you could eat and drink right from the jungle. What we discovered was that if you were lost in a jungle, it would be easier to die of old age than starvation.

That really impressed me.

One of the things that fascinated me was all the stuff that was edible in the jungle. We could drink fresh water from the vines that hung from the trees. We could eat fruit from a variety of trees and feast on tubers from the ground. Hack a banana tree down and you had food for weeks, plus an endless supply of fresh water that would refill the tree stump. I should not forget to mention the full red meat menu that prowled the jungle night and day (best to kill it before it killed you). Even the lethal snakes were alleged to taste like chicken if properly prepared. I never tested the theory.

When we weren't learning about what we could eat from the jungle buffet, our days and nights were filled with tactical exercises, initially at the squad, then the platoon, and finally the company level. The Army used a building block approach to the training that was professional and effective.

All the tactical training was interspersed with useful survival training.

We learned about the dangerous wild life, from the little fire ants that could swarm all over you and eat you alive, to the pumas that prowled their territory, to the snakes that could either strangle you (boa constrictors), swallow you whole (pythons), or kill you with one bite (the Fer de Lance).

On one of our first overnights in the jungle, before dark I was setting up my hammock between two trees, and my attention was drawn to the jungle floor where there was an eight lane highway, four each way, of ants carrying little slivers of green jungle leaf to wherever they were going.

During the second week we had some fun with both rappelling out of helicopters that hovered 120 feet above the ground, and being extracted from a (simulated) narrow jungle clearing by being lifted out on ropes dangling from said helicopters. We were in UH1E helicopters – the HUEY -- that was a mainstay in Vietnam. It had one single large rotor above the cockpit/passenger compartment, and a small vertical tail rotor. It is a really cool, dare I say the "Hot Rod" of military helicopters?

Our first experience with the concept of using helicopters in a Jungle Warfare environment came when we were to practice four-man extractions from a simulated hostile jungle environment.

We assembled in an open area and sorted ourselves into groups of four. Each of us strapped up in tight fitting harnesses that had steel rings on the shoulder straps. The officers went first, as was appropriate for all Marine training. At the appropriate time, our teams of four began the linking up process. First, the helicopters lowered the ropes, then each team linked a 'D' ring from each rope to the steel ring loops on each of our shoulder straps.

Once hooked up, each foursome linked arms and legs together and the HUEY lifted us off the ground. Soon the helicopter reached an altitude of about 600 feet, and there we were dangling about 120 feet below the helicopter as it took us on a leisurely sweeping tour of the Gulf of Panama above the eastern entrance to the Panama Canal.

You haven't seen the Panama Canal until you've seen it from 500 feet dangling below a helicopter.

After that bit of fun (and for me, at least, it was a lot of fun), we got to practice rappelling out of HUEYs at 120 feet. We had mastered the art of rappelling earlier in the week (and also at an earlier date back at Indiantown Gap) from twenty-five foot towers, but now it was a different matter. There would be no wall to bounce off of, and we would be doing it at a much greater altitude. Rappelling out of a HUEY at 120 feet was a quantum leap (pardon the pun) for us.

The four officers of the company – Capt. Vic Alt, Capt. Jim Shimp, Lt. Jack Kotanchik, and I were in the first group to demonstrate the technique (officers always set the example in the Marine Corps). The HUEY landed and we each took a position on one of the four corners of the troop compartment. Vic was on the left front corner, Jack on the left rear corner, Jim on the right front corner, and I was on the right rear corner. Our hook-up process had to be done in an exact way so that the rope would be looped through the 'D' ring in such a way that it would create a friction on itself to help slow the jumper's descent.

We dutifully hooked up, looped the rope through our "D' rings, were inspected by the Jumpmaster, and awaited the command from him to go. We all wore special reinforced gloves as the descent would cause a great friction on our hands, particularly the lower hand which was placed right below our buttocks. We dropped out of the aircraft in pairs. For balance of the aircraft, first off would be left front and right rear. Next would be right front and left rear.

I was in the first pair. In the right rear position after being checked out by the jumpmaster, I was given the green light to prepare for the jump. I moved out onto the right-side strut of the helicopter and leaned into the 90-degree jump-off position. On the command of the jumpmaster, Vic on the left front, and I on the right rear, simultaneously jumped off the helicopter struts. I began a controlled rappel 120 feet down to the ground. I was about 25 feet below the helicopter, holding my lower hand just below my rear end and periodically squeezing gently to slow my descent, when I saw this dark object go sailing past me at a high rate of speed. Jim Shimp had not been properly checked out by the Jumpmaster. His rope was looped improperly through the D-ring and was not creating the necessary friction to slow him down. As a result he was falling almost unchecked. Two things worked in his favor: one, he was a big, strong guy, so he clamped his lower gloved hand firmly around the rope to slow his fall; and two, there was what is called a "belay" team on the ground. The belay team immediately reacted when they saw he was in near free-fall and pulled hard on Jim's rope. While that action put pressure on the pilot to maintain altitude, it also served to help slow Jim's descent. While he had a hard landing, and his gloves were smoking, he lived to tell about it.

Later on, in another training event, we were assigned to our Huey helicopters in eight-man groups and commenced flying across the jungle terrain at tree-top height. This was the Army's concept of contour terrain flying. Rather than fly at a steady height of 800 or one thousand feet to reach the objective, the pilots would fly anywhere from fifty to one hundred feet of altitude while following the altitude contours of the terrain. Sitting sideways, four of us facing outwards, with boots on the struts while this was going on was a total rush.

The second week concluded with a three-day exercise in the jungle that built on all that we had learned. It was absolutely exhausting. As an aside, in the first week, we had learned how to string hammocks between trees so that the ants wouldn't carry us away in the dark. We had to learn how to adapt to that primeval environment. No way could we kill every fire ant, puma, or snake. We had to learn to survive alongside them.

One day early in the second week, my platoon had the lead as we navigated our way through the jungle en route to an assault objective. In trace behind us was the company command group (CO, XO, First Sgt., Corpsmen, et Administratively al). Following them was our other platoon. Our battalion masters had us halt while they figured out what direction they needed to go to next. It was a long pause, maybe an hour or so. I remember we were waiting for the word to come down as we sat along a trail by a river that had cut through the rocks maybe ten million years before. I expected dinosaurs to start roaming and pterodactyls to fly by. It looked that primitive. It was a scene right out of King Kong.

Finally battalion got word for us to move out. I passed the word back to move out, but someone at the end of my platoon forgot to pass the word back to the next man in the company command group behind us, and the result was that we inadvertently disconnected ourselves from our company headquarters and brother platoon. After about an hour we were ordered to halt as our company commander had managed to reach battalion by radio to indicate they were stranded. After they got the correct compass heading, they finally caught up with us. The commanding officer was tired and pissed, so he took it out on me. Later on we had a laugh and a few beers over it, so no permanent damage done.

Ultimately, at the conclusion of this exhausting two-week ordeal, we were given two days of liberty. A number of us checked out the Panama Canal, but the officers all decided to go across the isthmus to Panama City at the Pacific side of the canal. Boy was that a wild place. Truly an international city. One was just as likely to see an Arab as an Asian as a Panamanian. But the shopping was good and we got to sleep on clean sheets at the Holiday Inn for the first time in two weeks.

* * *

Then it was Saturday -- moving day -- and we packed up. Trucks took us to the Air Force base where we once again boarded the C-141s for the trip home. After all our gear was loaded we settled our 120 Marines into the web sling seats running forward to rear that hung in four rows along the length of the fuselage. I was sitting about four or five men from the ladder leading to the cockpit in a sling seat facing the port (left) side of the aircraft. Our First Sergeant, Glenn Laughman, was sitting opposite me with his back to the fuselage. We were both dog-tired, and I started to nod off as the plane taxied into takeoff position. There was the sudden roar and acceleration as this monster of an aircraft gained speed down the runway. Soon we were airborne and beginning to climb.

But then we stopped climbing and Air Force crew members were scrambling all over the plane. The First Sergeant and I locked eyes and read each other's thoughts. I stopped an airman who was passing in the aisle between me and the First Sergeant and asked what was wrong. Over the roar of the engines he shouted in my ear that the right landing gear had not fully retracted, nor would it fully go down into the landing position, it was canted at an angle. The pilot was circling the Gulf of Panama at three thousand feet and dumping fuel in preparation for making an emergency landing back in Panama.

After I relayed that information to 1st Sgt. Laughman, he and I locked eyes and both had the same thought. Our toughest training was going to end with our body parts strewn all over an airport runway in Panama. Write home and tell Ma.

Apparently the pilot's plan was to come in and touch down with the left hand landing gear then gradually let the plane settle on the right hand gear, which hopefully would snap it back into place. I suppose when it comes down to all the available options, short of heading for Mars, that one probably seemed the sanest. However, to my mind at the time, it seemed like a totally insane thing to do.

But I couldn't come up with an alternative.

We were all going to die.

Miracle of miracles, it worked. After our landing, we de-planed with all our gear and sacked out on the runway for a few hours until the maintenance crews thoroughly checked everything out and did the appropriate fixes. After all repairs and inspections were complete, it was back on the plane for eight hours, stopping in Charleston, SC, for fuel, then on to Harrisburg.

Finally home late that Sunday afternoon, I just wanted to sleep for a week, but I had to go back to work at Pennsylvania Blue Shield the next day. Suit and tie. Desk. Sell a computer system that would process health insurance claims. All very important in the overall effort to ensure that the wheels of the American economy were kept turning, but what a let down that was.

* * *

Hotel Company continued to thrive. In the spring of the following year we got a new I & I. Captain Bill Green, a North Carolina native with a drawl you could cut with a knife. He took over command of the I & I Staff, and before we had ever drilled with him being in charge of our support operations, he called a meeting with the company officers. He wanted to get a feel for our concerns and expectations about the training and I & I support.

I talked with the other officers, and we agreed to meet with Bill away from the reserve center at a place in Harrisburg, about a block from where I lived, called Abe's Tavern over on Third Street. At Abe's, we gave it to him with both barrels. We were tired of the disrespect the active duty staff showed to our enlisted, and we were outraged by the silent contempt with which they held the reserve officers.

Bill then said something that really stuck with us. And it became a leadership axiom for me. He said in his thick North Carolina drawl, "I dictate the attitude of the staff, and you will see positive change immediately."

He was good to his word and a most cooperative atmosphere prevailed for his entire tour of duty. His First Sergeant, who had been the biggest prick of all, suddenly became the soul of cooperation and respect. The rest of the staff followed suit.

Monthly progress meetings at Abe's became the norm.

One of the best things Bill did was create a strong liaison with the Army and National Guard units at our field training site, Indiantown Gap, PA. Located about twenty-two miles north of Harrisburg, the Gap was host to squadrons of Pennsylvania National Guard helicopters; had major live firing ranges for rifle fire, artillery, mortars, and machine guns; a live-fire pop-up target range for squad fire and maneuver exercises; and also had many open and forested training areas for tactical maneuver training. It hosted an Armored Personnel Carrier Army Reserve unit that rarely trained there. It also had a Learning Center, lightly used by the Army Reserve or National Guard units, where the troops and NCOs could learn the basics of fire support in a classroom environment. That means nothing to you but was huge for us.

Best of all, Bill Green and the Director of the Learning Center had been classmates at the Advanced Infantry School at Fort Benning, GA. What made the connection special was that Bill regarded this Army officer as a close friend. Bill was White North Carolina, The Army officer was Black Alabama.

Is this a great country or what?

Under the previous I & I, we only went to the field about four times per year. Thanks to Bill's aggressive liaison efforts with his Army friend, for the next several years we were in the field eleven weekends out of twelve every year. The twelfth weekend was needed in garrison to conduct necessary administrative activities.

Bill got us great support. It seemed that the Army Reserve and National Guard units who were supposed to use all these vehicles and ammunition and helicopters and firing ranges had little interest in doing so.

It turned out that there were tons of Army fighting vehicles that we were able to use. Bill arranged to get a bunch of our troops licensed to drive them. As a result, we were able to simulate Marine AMTRAK maneuvers with them.

The helicopter squadrons were delighted to do tactical insertion and extraction operations using real troops (many of the pilots having been Marine pilots in Vietnam who had opted to continue Reserve service with the Army Reserve or National Guard just so they could continue flying). Our machine gun and mortar sections were given unlimited Army ammo and conducted live fire exercises every drill weekend; and there was also a squad live fire and maneuver pop-up target range where our troops could get tactical training under realistic simulated live-fire combat conditions.

Every drill weekend we split the company into sections and had rotational training going on constantly.

The Army basically gave us carte blanche because their own reserve and Guard units were not coming out to use up all these goodies. And a basic rule of military funding is that if you don't use it, you lose it. So the more bullets we could fire, the more the Army could order for next year.

One other aspect of this that Bill helped with was the maintenance of our weapons. It is not enough to just clean weapons after firing them when you return to the reserve center. If you have been doing a lot of live fire, you also need to put each weapon through what is known as a TMI (Tactical Military Inspection), which is an intense inspection and re-calibration of the weapon. On Active Duty this type of inspection might happen once every six months. We had no such capability in Harrisburg, but there was an Army Reserve unit near Philadelphia that did nothing else. Bill arranged for our weapons to be shipped there each month the Monday after we got back from the field. So Sunday afternoons of drill weekends we would clean and inspect the weapons, and Monday morning they would be shipped out to Philadelphia for further inspection and re-calibration. Our weapons were returned within two weeks in plenty of time for our next training weekend. There's not a Marine Corps Reserve unit that could match this capability.

The bottom line was that when we got to our annual training, we blew our active duty counterparts away with both the condition of our weapons and our proficiency with them. Most of the active duty units could not match what we did monthly.

Our training progressed, and our NCOs, troops, and officers became more proficient.

Unfortunately, we had a major tragedy at our Annual Training in 1977.

The Marine Corps chartered a Boeing 727 to take us from Harrisburg to Coronado, California, for the first stage of our annual training that year. We landed in Salinas, KS, to refuel shortly after major thunderstorms had raged through the area. The weather was awful with lots of violent thunderstorms in the area. This was, of course, the heart of the US's "Tornado Alley." After we had re-fueled, we waited another hour or so for a fair weather clearance, then re-embarked aboard the 727 and began the process of take-off.

As I have indicated, it was very stormy in the region. The airport, having been built during World War II with lengthy runways that accommodated B-29 bombers, had, prior to our landing, been saturated with major thunderstorms. The runway was not graded as today's modern runways are so that there are no dips to create puddles. It was an old, concrete runway.

We taxied to position and then began accelerating to take-off speed. We were splashing through some huge puddles of water. The 727 aircraft has its engines mounted up on either side of its tail, and its wheels are located under the wings. In inclement weather such as we had been experiencing, the jet engines were subjected to the negative effects of water puddles that the plane's wheels have splashed through. Those splashes forced water right up into the intakes of the jet engines.

Sure enough, as we approached lift-off speed, there was the loudest bang I never want to hear again. Water had splashed up from the wheels into the intakes of the jets on either side of the tail of this 727. First Sergeant Laughlin, who was seated next to me, and I turned to each other, locked eyes, and we knew that within seconds, our lives were to end. (Didn't we just do this in Panama last year?)

Not so. The pilot came on, and in his best Chuck Yeager imitation (see "The Right Stuff" movie or book by Tom Wolff), and drawled, "Well, we had a bit of compressor stall, but we're just going to taxi around and give it another shot."

Compressor stall, for those of you flyers who may suddenly hear the loudest bang that convinces you that the end of your life is imminent, is an aeronautical phenomena where the jet engine intake system is no longer receiving the air it needs to continue functioning. That, my friends, is the exact point where your pilot loses airspeed, altitude, and experience, all at once.

Thank you for flying with us today.

We returned to our start point and began the process of defying gravity once again.

Somehow we survived Salinas, Kansas.

We flew to San Diego, CA, and then took a short truck ride to Coronado for amphibious training. We arrived late afternoon on a Saturday and starting on Sunday, conducted rigorous training that simulated disembarking from amphibious ships in combat situations.

One of the realistic aspects of the training program at Coronado was Survival Training. We went out about a mile off-shore in DUKWs (Duck Boats that are now used in many cities to conduct waterborne tours – these were wheeled amphibious vehicles). We Marines were in life vests but also carried our combat gear – rifles (old rusted ones used only for this training), our belt suspenders with magazines, ammo pouches and first aid packs. For the survival program, we walked off the bow two at a time as the DUKW slowly backed up. We did this one squad at a time. Well Sgt. Webb, one of my up-and-coming squad leaders, was terrified of water. So here I had Sgt. Weikert terrified of heights and Sgt. Webb terrified of water. I gave Sgt. Webb a similar pep talk like I had given Sgt. Weikert at the "slide-for-life" event back in Pennsylvania. You want to be a Leader of Marines, you have to set the example. I told him to hold onto my hand and relax.

We were the first pair who stepped off the DUKW and hit the cold Pacific Ocean. As one born and raised on the Atlantic coast with the influence of the Gulf Stream and the long Atlantic shelf, I was totally unprepared. I mean this was early August, but the Pacific is one cold freakin' ocean compared to the Atlantic. The Gulf Stream in the Atlantic, coupled with the not too deep long Atlantic shelf extending two-to three-hundred miles off the Atlantic coast keeps the swimming waters of that ocean at sensible temperatures. Generally coastal Rhode Island can anticipate water temperatures to be in the mid- to high-seventies in early August. The Pacific had no such protection. The waters off-shore quickly dropped off to a greater depth than the Atlantic coast, and there was no warming Gulf Stream. We probably hit the waters at a cool mid- to high-60s temperature. That may be generous of me and was shocking to all of us.

Once we stepped into this stunning cold, Sgt. Webb showed the beginnings panic, but quickly calmed himself. I gripped his hand and told him to have faith in his gear and his fellow Marines. He put his trust in me and I could feel his grip relax. We gradually hooked up with the rest of the platoon, hooked arms, and back-pedaled our way to the beach. Sgt. Webb had acquitted himself well in the ocean, as Sgt. Weikert had acquitted himself on the slide for life tower, and once again validated the trust I had placed in both of them.

Further amphibious training included embarking and disembarking ships via cargo nets strung over the sides (no longer done in Marine amphibious training as both embarkation and debarkation are now conducted by entering amphibious craft from within the holds of Navy shipping). But back then, our amphibious landing approach was driven by the capabilities of much older amphibious ships.

After climbing down those rope ladders that our predecessors had descended in WWII, I developed a new appreciation for Marine amphibious operations. The ships bobbed up and down as you, in full combat gear, descended into the "Mike" boats pulled up alongside. The ship was going up and down, the boat was going up and down, and neither was in sync with the other. The last step off the rope ladder was usually nothing more than a leap of faith that your fellow Marines would catch you.

And they always did.

On Thursday we boarded Navy amphibious ships for a short run up the coast to Camp Pendleton where we conducted an amphibious landing in a hurricane. These Marines had only a few days to absorb amphibious training with only one shot at an actual amphibious landing. This seemed to be standard protocol for the Marines. Do it, weather be damned.

At our landing at Pendleton during this horrendous storm, we all thought, from the time we disembarked the ship, that we were going to die. All too soon our landing craft got hung up on a sandbar near the beach. The coxswain, that is the Navy Petty Officer who drove our landing craft ashore, lowered the bow gate while gunning the engine to stabilize the landing craft on the sandbar, and we stepped off into six-foot deep surf. Fortunately Navy Beach Masters were there to drag us to shore. Somehow we got everyone onto land (not dry, but alive) with no casualties.

Once on the beach it was the standard "deploy for fire and maneuver" to take a series of objectives. We stayed in the field for two days (mostly soaked), then were trucked over to some really horrendous barracks that had been abandoned at least a year earlier. We spent a dry but grossed out night in this temporary housing that should have been declared a waste dump by the Department of Environmental Management. These buildings should have been napalmed. Thank God we were there for only one night. The accommodations were unfit for human habitation.

The Marine Corps should have been ashamed to house any of its troops in such conditions. A fundamental rule of leadership is to take care of the troops. You can't do that in sub-standard housing conditions. But, as I mentioned earlier, there was still a huge 'Us versus Them' attitude regarding how the Active Force treated its Reserve component.

We were the casualties of institutional indifference. That was not to change until some years later.

The following morning, Sunday, it was bright and sunny and we were trucked to an airfield where we boarded C-130 transports for a flight over the mountains to Twenty-Nine Palms, California, the Marines' desert warfare center.

After a few days of training in Coronado's beautiful summer weather, landing in a hurricane at Camp Pendleton and being chilled to the bone, sleeping in inadequate barracks, and then being flown over the mountains to the high Mojave desert, I remember getting off the plane at 29 Palms, looking around, and being stunned at the geographical and topographical differences. I remember making some comment that it looked like a moonscape. This was the high Mojave Desert, and it was, in reality, a moonscape. None of us had seen anything so desolate and (presumably) empty of life in our lives.

When we de-planed on that late morning, we were all astounded by the desert heat. During daytime in the high Mojave desert in early August, the temperature regularly hit 115 degrees.

We were brought by truck to a tent camp named Camp Wilson. After disembarking and getting the troops settled, we got our briefing. The following morning we were to pack up and move to the middle of nowhere in the desert and get ready to launch an attack (on God knows what).

After getting settled in our tent camp, we faced the next shock. The nighttime temperature dropped to 65 degrees. You're sitting there saying, well, duh, that's not so bad, and I'm telling you that after dealing with 115 degree dry desert heat, to have the temperature drop fifty-degrees in a few hours was like being stranded on New Hampshire's Mount Washington in January.

Thank God we had really comfortable sleeping bags to ward off the chill. But before we turned in, and each night we were in the desert, I looked up into the crisp, clear night sky and saw for the first time in my life, the Milky

Way Galaxy where we, in a little corner of that great swirl, live and try to thrive. Seeing those billions of stars was a moving experience. And the shooting star experience every night, particularly in early August, was spectacular.

Our wonderful Hotel Company did its part and the battalion staff screwed up theirs. We sat all day out in the desert awaiting orders to jump off in the attack, but there was always something delaying the order. That first evening in the desert our Weapons Platoon decided to do some morale building. The unit cleared a circle in the sand about twenty-five feet across and started having wrestling challenges. At first they were content to keep it within their platoon, but then someone spotted me and challenged me to come in and wrestle one of their troops. A product of the New England school system, I had never wrestled as most of these kids did in high school. Wrestling was just not part of our academic athletic experience. In Pennsylvania, however, it is a major school sport. But, as I firmly believed, an officer does what he or she can to improve morale.

I went into the circle and looked over my possible opponents. This was a fun bit of theater. Standing off to the side with his usual scowl on his face, massive arms folded across his chest, was Private Foster. Foster stood about six foot four inches tall and was built like somebody's statue. I mean he was jacked. And he always had a mean look on his face. If you look back at boxing footage from the nineteen-sixties, you would see that Marine Private Foster had the same scowl and hatred on his face that Sonny Liston had before he was beaten by Muhammed Ali. Foster was a problem Marine because he was dealing with a lot of Black anger (in no short supply in the military of the 70s), and he was dealing with perceived slights to him and the other Black Marines. It was one of the legacies of the seventies when racial tension was high, even though our H Company officers had bent over backward to make sure that every Marine in our company was treated the same.

So I make a quick decision and choose Foster as my opponent.

And the crowd, as they say, went wild.

Dear reader, I will shorten the agony you would suffer from a full-blown narrative of the event. Suffice it to say that young Private Foster trounced Captain Power handily, thank you.

To his great credit, and with a rare smile, he offered me a hand as I wobbled to my feet. I felt lucky to be alive. This guy was one hulking human.

So now, as the defeated participant, I got to choose my next victim. I called for Sgt. Weikert to perform the next challenge. As he had once been a high school champion wrestler, I thought he would redeem my poor performance. I forget who he chose to fight, but he acquitted himself well. However, after the match as he was leaving the circle, he looked at me and said, "Skipper, I'm not feeling well."

Horrified by his sickly demeanor, I made him sit down and called for a Corpsman who, fortunately, was nearby. The Corpsman asked Weikert what his symptoms were. Gary responded that he was having chest pain, soreness in his left jaw, and pain in his left arm. The Doc looked at me and I said, "Medevac him now."

They quickly got Gary to the battalion aid station and he seemed to be feeling a little better, then his heart just gave out and he died of a massive myocardial enfarction on the exam table with a doctor standing right next to him. The autopsy later revealed that this twenty-five year old Marine, a physical specimen with the physique of an athlete, had been born with a diseased heart that got worse year by year.

When, around dawn, the Gunnery Sergeant with the I & I team woke me to tell me that Sgt. Weikert had passed away during the night, I lost it. I crawled back into my little shelter half and bawled like a baby. The good Gunny stood guard over me and wouldn't let anyone near my tent. Jeepers, here it is forty-five years later and it just dawned on me that I was notified first. The system dictated that the Commanding Officer should have been notified first. But they notified me. Oh my.

Finally, after I managed to recover, I left my tent, went to Vic's (the CO) and told him of Gary's passing and that we needed to continue the training. If we stayed where we were and wallowed in Gary's death, we would never recover. Vic agreed and asked me to announce that to the troops. It was his job to do it, but I did it.

Vic, our CO, didn't want the responsibility of telling the troops that we must continue the training. As his XO, the duty, and the negative fall-out, fell to me. I had several troops as well as one platoon sergeant accuse me of being heartless. The problem was that my heart was breaking because of the bond I had established with Gary Weikert. But there was another part of me, the professional Marine officer, that said to himself, "Don't let the troops get bogged down in sorrow and self-pity. We have a mission to accomplish."

I loved Sergeant Gary Weikert. I was the one who chose him to become a leader of Marines. I mentored him. He led troops in protecting his Captain's home from the devastation of flood. His loyalty to me was beyond what I can describe. I was bereft.

I think I was ultimately forgiven. Even if I wasn't, I know I made the right decision.

Vic asked me to brief the battalion commander on the situation (another example of 'it was his job but I had to do it'), so I took First Sgt. Laughman with me and we briefed the Colonel, his S-3 (Operations), and his S-1 (Personnel) officers. They agreed that we needed to keep the troops moving and as motivated as we could.

I ordered Sgt. Ted Webb, one of Gary's best friends and fellow squad leader (and the other Marine I had identified as a future leader), to accompany Gary's body back to Pennsylvania. Before he departed, the I & I staff gave him a briefing on his duties. After arriving in Harrisburg and getting the information he needed from the funeral home, Sgt. Webb drove down to his home in Chambersburg, PA, to retrieve his Dress Blue uniform. For the next five-days this wonderful Black Marine never left Gary's side except to make a head call or take a quick shower. By the time the company got back to Harrisburg, we had a brief time to clean up, change gear, and report in dress uniforms for Sgt. Weikert's funeral. It was beautifully done with the full honors that only the Marine Corps can provide, and we sent our brother home. Though I was personally totally crushed, I could not show that to the troops. And that is the main challenge of leadership.

In the meantime, back at 29 Palms, we went back to our bivouac area and got ready to move out in the attack.

But first, the battalion staff had to unscrew itself. It took until late the following afternoon before they were ready to execute battle orders. While we were waiting for that to happen, I was sitting on some high ground inside a combined lean-to we had patched up by hooking four shelter-halves together. With me in this primitive enclave were Capt. Vic Alt, Lt. Jack Kotanchik, and First Sgt. Glenn Laughman.

We looked out across the desert and we all became alarmed at the same time. Several miles away to our right, there was a major storm and it clearly was moving our way. What further alarmed us was that as we looked out in front of us and also down to our left, there were two other major storms.

Within a few minutes you could see that all three of them were going to converge on our position.

These storms, interestingly, sounded like the storms that happened in Hawaii after the passing of a great 'Ali Nui' leader. I wonder if Gary's spirit had anything to do with it.

We alerted our troops to gather up all loose gear and button themselves up in their two-man shelter-half tents to ride out the storm. Everyone turned to, picked up all loose gear and secured themselves as best they could. Shortly the first storm arrived and it was mostly wind. I'm talking about serious 70 mile per hour winds, maybe higher.

Just after it passed, the second storm hit, also with high winds, and this storm was mainly hail. We were all wearing our helmets and flak jackets. Even though we were in our shelters, the hail beat down the canvas and you could hear it hitting helmets all over the place.

That storm moved on and was replaced with the third storm that dropped a Biblical level of rain on our camp. It rained so hard, and over such a short time-frame, that the next thing we had to worry about was flash-flooding. The desert floor is baked to a hard-packed surface most of the year. When it does rain, the rain doesn't seep down. So the only recourse for the water is to flow down and run off any crevasse available.

We had established our camp along what had appeared to be a dry wadi. When this third storm hit, the water just started to rampage down any slope available. Fortunately all of our troops, save two, had camped above the wadi on level ground. Two of our guys had pitched their tent on the slope of the wadi. And suddenly the water was rising at a furious rate.

We officers had pitched our shelter on the high ground across the wadi from these two Marines and I could see that because they had buttoned up their tent, they had no way of realizing the danger they were in. The water, as I said, was rising at an alarming rate, so I just ran out of our shelter, struggled across the raging torrent, and got to the other side just in time to rip their tent open and drag both of them to safety. Within seconds the torrent tore their tent away and dragged all their gear and personal possessions downstream, probably to end up somewhere in Arizona.

The next several hours were spent in straightening out our area.

Finally the order came down and we were to move out. The sun was back up in the sky doing its job of baking us to a crisp. However, we had been briefed on a survival technique that was taught to the Marine Corps by the Israeli Army.

The Israelis, constantly besieged by their surrounding Arab enemies, had, in both 1967 and 1976, having been attacked by five Arab armies, decisively defeated their foes. One of the ways they did it was doing what we ended up calling "overdrink." In both of these wars, the Israelis suffered very few heat casualties. But in the 1976 war, the Egyptians alone suffered 20,000 heat casualties.

The solution was two-fold. First, every soldier (or Marine) in the desert must drink four to six-gallons of water per day; and second, whether he was hungry or not, his squad leader must visually confirm that each of his troops had at least something to eat at each meal.

So each of our squad leaders, every half-hour, made sure that each of his Marines drank at least one-half canteen of water (fortunately our water resupply was endless), and that at each mealtime, each of his Marines had something to eat.

We had a few casualties that summer, but none from the heat. Mostly it was pulled muscles or sprained ankles and the like.

* * *

Over the next two years that I was affiliated with Hotel Company, we added Captain Doug O'Dell as the new Weapons Platoon commander, and Captain Ray Hedger to replace Jim Shimp as Second Platoon Commander. Doug O'Dell was to distinguish himself by rising through the officer ranks to become a Major General (two-stars) before retiring.

I spent two more years with Hotel Company before I got promoted to Major, and then I had too much rank to stay. Before that happened, I had an opportunity to assume command of the company. Vic Alt was finishing his two-year tour as Commanding Officer, and I was the heir apparent. But then, as it always does, fate stepped in. The battalion commander told Vic that there was another Captain, highly regarded and slightly senior to me, who was in another rifle company in the battalion and, because of a three-year broken time between active service and joining the reserves, was faced with a "get command or get out" dilemma. Even though he had the option of refusal and could have insisted on me taking over the company (because I had goddamned well earned it!), Vic caved in to the battalion level pressure.

Command was not to be mine. For the past two years it had been mine in practice, but not in fact.

Captain Bill Loop came in as the new Commanding Officer, and he was a good guy, professional, competent, and did a fine job. We all, myself included, liked him, and did our best to keep our company at peak readiness. I stayed with the company another year or so before my own promotion to Major made Company Command a moot point for me. The gods of chance had spoken. It was time to move on.

During my time in Hotel Company, I had the great pleasure of serving with (then) First Sergeant Laughman, who was one of the most competent senior NCOs I ever met (not to mention one of the funniest men I ever knew), and who, a short time after I left the company, was promoted – justifiably – to the rank of Sergeant Major; and officers like the aforementioned Jack Kotanchik, and fellow captain Doug O'Dell. When I first met Doug he was in the financial advisor business. He later moved to Texas and made millions in the oil business. He had a quick and sophisticated wit, an amazing gift for leadership, and the troops loved him. Also in the company, replacing Jim Shimp who had moved on to another assignment, was Captain Ray Hedger. Ray was an extremely successful tax attorney, and a very competent Marine officer.

One of the more interesting sidelights to the varied training experiences we had undergone was how it motivated a wide range of troops. Sgt. Strawser really wanted nothing more than to be back on active duty in the Marine Corps. I think his wife was reluctant, but after the three years we had completed with Amphibious Landings and the Raid Exercise at LeJeune; Jungle Warfare School in Panama; and Desert Training in the high Mojave under our belts, civilian life paled for him. His wife finally agreed and he re-enlisted and went back on active duty.

Similarly, Sgt. Ted Webb, with a well-paying long-haul trucker career ahead of him, also decided that civilian life was too tame, and he returned to active duty. Unfortunately, twenty years later, I read his obituary in one of my Marine publications. I did some research and found that his mother was still alive. I wrote her a letter praising Ted's leadership skills, and particularly his service to Sgt. Weikert and his family, and she was kind enough to send me a lovely response thanking me.

We also had a Sgt. Enterline in Hotel Company. He was in Jim Shimp's platoon and had been on active duty for five or six years before getting out and joining the Reserves. Following our Jungle Warfare School and Desert training, he also was so motivated that he returned to active duty.

All in all, there were at least five Marines from Hotel Company that went back on active duty. It was unprecedented, but I think it was a reflection of the leadership Jim Shimp, Jack Kotanchik, Doug O'Dell, Ray Hedger, Bill Loop, and I were able to provide.

In closing this section, I recall a hilarious moment one weekend when we were training out at Indiantown Gap and it started pouring rain. I mean it was coming down in buckets. Well, there we were, just having finished training for the day on Saturday, and we were not scheduled for overnighting in barracks. We would be in tents that were already flooded. The trucks carrying our hot meals in big vats pulled up to the training site, and the company dutifully lined up, everyone in rain poncho, to get their mess kits filled with what immediately would be hot but soggy chow. The officers, of course, waited until all the troops were served, then we got our mess kits out, lined up, and got our share of the chow.

There was no shelter, so we walked several paces away from the trucks and sat down in a circle in the mud, rain pouring down in buckets, and began to eat our meal. Very quietly, Capt. Doug O'Dell asked our circle, "Why are two health insurance company managers, a tax lawyer, the owner of a big lumber retailer, and a stock broker out here getting soaked and eating a soggy meal?"

Jack Kotanchek responded with the same thing he said to me in Panama, "I'm only in it for the money."

Well I thought the lot of us would choke on the soggy mess in front of us from all the laughter. It was one of those uniquely Marine Corps moments that I think only Marines in the field can fully appreciate.

14.

The MTU (Redux) and Wargaming

As a Major I had now become too senior in rank to remain with Hotel Company, I moved on for a year once again with a Volunteer (i.e., no pay, but with earned Reserve retirement points, and later designated "Mobilization") Training Unit that specialized in war games. This unit was loaded with terrific mid-career Marine Reserve officers. Like many reserve units that didn't fit into the standard vision of fighting Marines, it was a one-of-a-kind and unsung institutional hero of my Reserve world. It was an unpaid think tank.

Before I had joined this MTU (a unit populated with many of the same Marine officers I had originally met when I was a student at Mount St. Mary's College when I had joined the VTU in Harrisburg), it had spent the past year perfecting a "table" war game simulating a Soviet Union attack on the plains of western Germany. This game was special. It had one-foot square sections of molded terrain that approximated the terrain of western Germany. The unique feature of this model was that it could be reconfigured with the other one-foot square sections in an endless rolling display of battlefield terrain.

The game, as most of those pre-computer games were designed, was driven by the throw of a twenty-sided dice. This unit had spent countless hours, well beyond what the Marine Corps would have granted as legitimate Reserve time for Retirement points, in perfecting this model.

But after close to two years of testing this model, it was ready for a serious road test.

We carefully packed up the sections of the war game and shipped them to Orlando, Florida, where a team of moderators, controllers, and observers would put the game through its paces. The people who were to "road-test" the game were all retired Marine officers.

Headquarters Marine Corps had cut individual orders for each of us in the MTU to go down to Orlando for two weeks to test the game.

It turned out that we had produced a pretty complete game that could be replicated and War-Gamed in both Army and Marine schools.

The Commanding Officer of our unit would receive a Navy Commendation medal for leading this effort. The officers who put in the time creating this model, of course, got nothing.

15.

Reserve Division HQ Marine Corps, Reserve Augmentation Unit (RAU)

1981-1984

Somewhere around 1981 (for those keeping score), after a year with the MTU, I discovered another paid opportunity in the Marine Corps Reserve. I applied for and was accepted into the Reserve Division Reserve Augmentation Unit (RAU) at Headquarters Marine Corps (HQMC). Whenever a Marine was attached to a RAU, it meant that he or she was on individual (vice unit) orders. But, unlike the MTU's, the RAUs were paid positions.

This was an interesting opportunity. This Particular RAU had just been formed and the officer organizing it had a great deal of latitude in giving the RAU members extra days of Active Duty in addition to our monthly Reserve two-days of obligation.

Being a salesman on commission (I was then working in sales at Dictaphone Corporation), I could make my own hours and my own schedule. As the Marine Corps Reserve was proving to be much more interesting than any of my attempts to succeed in the world of capitalism, I jumped at the opportunity to pick up both extra Active Duty days with Reserve Division RAU as well as any additional two-week Annual Training assignments I could discover.

I spent about three years (through July of 1984) at HQMC commuting from Harrisburg to DC (and staying with my in-laws in Arlington, VA), and picking up as many active duty days, in addition to the mandatory two days per month, that I could. I also began requesting, and getting, as many extra two week training opportunities as I could. Most of these were schools.

In addition to doing two-weeks annual unit training on active duty at Headquarters Marine Corps, I managed to get an additional two weeks in 1981 on a special project at Quantico with a team of Active Duty officers who were in the final stages of bringing the Light Armored Vehicle (LAV) project to its completion. This team of a dozen or so mid-career (Major through Colonel) officers had been thrown together to research the plausibility of creating a light armored vehicle that could be transported by helicopter or driven to shore off of landing ships. The purpose of this vehicle was to put a six-man crew aboard and drive it deep into enemy territory to do reconnaissance missions.

That was a classic assignment. The Marines realized that upon breaching a hostile beach in an amphibious assault, under modern battlefield conditions, the assault force would need a rapidly moving tactical reconnaissance unit that could probe in-depth well into enemy lines and extract itself quickly. Such a unit would need a light armored vehicle to carry a small reconnaissance team.

The vehicle had to be able to withstand ordinary ground fire from rifles and must also pack an offensive punch with a small canon of its own mounted on a turret, much like a larger version that one could see on a tank.

The vehicle must be light enough to be lifted from ship to shore in an amphibious assault by a CH-53 helicopter. No helicopter being available, the vehicle must also be capable of leaving a ship and 'driving' itself ashore in surf conditions of less than four feet.

The Marines had no vehicle in the inventory that could do the job (nor did the Army), and the Army, from which we purchased all our major tactical fighting vehicles, also had no interest in pursuing that line of acquisition. The Marine Corps moved forward on its own, with the blessing of the US Congress. The Marine Corps planners now had twenty-four months to complete its task to successfully develop a program that would produce such a vehicle. And that vehicle would be unique to the Corps.

The Army laughed and said it couldn't be done.

The Army flat out told the Marines that what the Corps wanted was something that generally took ten-years to produce. The Army wanted nothing to do with it. But the Marine Corps deemed this future vehicle to be a strategic, as well as tactical, necessity.

Despite the Army's attitude, Congress, on the other hand, was convinced by the Marine Corps argument. It granted the Marines exploratory seed money to pursue the feasibility of the concept, and the Corps ran with it. The Corps solicited top rated infantry officers from each command (the three Marine infantry divisions) to staff the project. Just as I indicated in my earlier retelling of my experience in Iwakuni, Japan, when assembling an honor guard for the Commandant's annual inspection, the operating units used this new opportunity to get rid of, at least temporarily, what they perceived to be their officers who couldn't seem to get along with the program. So with no collaboration of any kind, units from all over the Marine Corps decided to 'unload' those officers who had that unseemly tendency to be, God forbid, "independent thinkers!!!"

As a result, by mid-year 1979 at Quantico, Virginia, a team of twelve officers, all of whom were condemned for coloring outside the lines, and all deemed pains in the ass by their former parent commands, and all passed-over for promotion, assembled to see if a light armored reconnaissance vehicle was a feasible option for the Marine Corps.

The Corps couldn't have chosen a better crew if they tried. These guys were exceptional. They were independent thinkers, unafraid of trying something new.

For two years the team worked on this new offensive concept. With the aid of a 'Beltway Bandit' consulting group (the common appellative for the government contractors that seem to suck up so much of the Defense budget and whose offices were strategically placed within the beltway surrounding Washington, DC), the LAV team developed a clear, step-by-step approach to answering the question of whether this inventive concept was feasible. The team sent out RFPs (Requests For Proposals) to the defense contractor industries to see if there was an "off-the-shelf" product that would meet, or come close to meeting, the Marine Corps needs.

Four contractors offered vehicles in production that could meet most of the basic requirements. All that remained to be done was to field test each vehicle. By the time I arrived for my two-week annual training as an additional administrative officer attached to this project, the final results were coming in, and the project team was bringing everything to a conclusion. As I absorbed the initial briefing, I was thrilled by what I heard. To give an example, one Marine Major, well-decorated with a Silver Star, Bronze Star (both for combat bravery), and also a Purple Heart for wounds received in combat, solved a particularly thorny problem with a little creative thinking.

He was running one of the finalist vehicles through its paces during testing operations at the Yuma Proving Grounds in Arizona when the vehicle experienced a minor breakage. One of the rungs needed for troops to climb in and out of the vehicle via the upper turret hatchway broke off in a training accident. The Major took the vehicle to the repair facility on base at the Yuma proving grounds to see how long it would take to fix it. He was told it would be six- to eight-weeks.

He thought to himself that such a delay for such a minor problem was not only incomprehensible, but just plain stupid. So he loaded the crew and its defense contractor observers onto the vehicle and told the driver to drive the vehicle out the main gate to the public highway and keep moving until he saw an automotive body repair shop.

Soon they came upon an appropriate repair facility. The people at the shop were duly surprised to have a military field vehicle drive in. When the Major explained the problem and asked both how long a repair would take and how much it would cost, the owner of the facility said that it would take ten minutes and cost about $15. The Major got out his wallet, sealed the deal, and soon they were on their way.

The whole LAV project worked that way. The basic mantra was to find the quick solution, take initiative, and make it happen.

To make a long story short, the Marine team I had been attached to, within two-years of having been assigned the project, delivered a final recommendation to Congress to make a purchase of the LAV for the Marine Corps.

The Army cried "foul" for no other reason than that such a process should have taken ten years for them. To the Army, that was how procurement of major vehicle systems worked. If the Marines could do the job in 24 months, Congress might expect the same results from the Army.

The team of Marine officer rejects did the job in 24-months, and I was lucky enough to be in on the tail end of it. As an aside, the team of Marine Corps officers, more or less rejected from their former commands for being a little too creative, and who pulled off this miracle, were all highly decorated (Silver Stars, Bronze Stars, Purple Hearts) veterans of Vietnam. They just saw no point in painting by the numbers.

* * *

I had a number of additional two-week Reserve training assignments including two weeks at the Landing Forces Staff Planning School at Norfolk. The following year I did the second increment of that school at Coronado, CA.

I also managed to squeeze in the two-week Policy and Strategy course at the Naval War College in Newport in February of 1982, which required an extensive preparation involving a reading of the *Peloponnesian Wars* by Thucydides, and the writing of an analysis of the tome. The Peloponnesian Wars took place 2400 years ago.

My typical reserve year during the 1981 through 1984 period included four to six weeks of additional training each year, and it really paid off.

At the Reserve Division Reserve Augmentation Unit in DC, I had opportunities in 1983 and 1984 to work as a liaison officer on two separate Global War Games. The first, in 1983, was at the Pentagon, the other was in the spring of 1984 at the National War College on the Fort McNair (Army) campus in Washington, DC.

While both were very interesting, the Fort McNair game afforded me the opportunity to interact with our NATO colleagues. I was assigned to provide administrative and technical support to the NATO cell that was playing the War Game. Commanding the cell was a Belgian Air Force Lieutenant Colonel. He and I had many interesting conversations, but one of them stuck with me. Remember, this was 1983 at the height of the Cold War when we were momentarily expecting the Russians to invade western Europe. I asked him, "Colonel, what exactly is going to happen if the Russians do invade western Europe?"

He responded in a somewhat droll fashion, and in a charming French accent, "We are prepared to defend western Europe to the last German."

We were, of course, only a generation or so removed from memory of World War II.

I managed in those two war games to get two "water walker" fitness reports signed by three-star generals. I also, in the Pentagon war game, got a letter of commendation from the Assistant Secretary of Defense for Reserve Affairs. I was very proud of that. This prepped me very well for the next major move in my reserve career.

In the spring of 1984, shortly after completing the Global War Game at Fort McNair, I received my acceptance as one of only two Marine Corps Reserve officers for the one-year residency program at the Naval War College in Newport, RI for the 1984 – 1985 school year. Upon successful conclusion, I would receive a masters degree in International Relations. On top of that, in late spring of 1984, I was selected for promotion to Lieutenant Colonel.

Now that was a good year.

16.

The Naval War College, Newport, Rhode Island

1984-1985

After much discussion about whether I would go to Newport as a geographic bachelor and come home on selected weekends, we decided to sell the house in Harrisburg, and move to Rhode Island. Our decision was a temporary one. We felt that the Naval War College opportunity was unique. Kathy said that while we had lived in Harrisburg for ten years, that was the longest she had lived anywhere. And as attached in many ways as I felt to Harrisburg, the lure of the New England coast was stronger.

I was born and raised in southern Connecticut right on Long Island Sound and grew up with the smell of salt-water and the fragrance of changing tides. Kathy's dad, a career Navy officer, had been stationed as the engineer officer at the Charlestown Navy Yard in Boston for three years. She loved New England.

We put our house on the market, the movers came and packed us up, and we were off to a new adventure. We knew that this new adventure was for one year. We had no plan or idea for what might follow.

The boys were not too happy about our decision, but they were young enough to make the adjustment. It was a strain for all of us because, in our ten years in Harrisburg, we had set in some roots, got really involved with our parish, met new friends, and lived in a terrific mixed neighborhood.

But, despite all of our doubts, we put Harrisburg in our rearview mirror and headed north. We checked in to the base at Newport and were assigned military student housing.

Newport is the southernmost community on Aquidneck Island which sits at the mouth of Narragansett Bay. Aquidneck Island is composed of three incorporated cities: Portsmouth in the north, Middletown, and then Newport on the southern end of the island. Our Naval housing was in Middletown. And there we began what was to turn out to be one absolutely fabulous year.

Shortly after we arrived the War College hosted a picnic/field day for all the new families. We dutifully showed up and were starting to meet new friends, when suddenly a football landed right next to us. Kathy picked it up and saw that a young teenager was coming toward her to get the ball back. She motioned to the young man to go out long as a quarterback would motion his ends, and the boy took a few steps back. Kathy then motioned more vehemently to go long. The boy shrugged his shoulders and started jogging away. Again she motioned him to go deeper, so he just figured, "I'll show this officer's wife that she is in over her head."

He took off and Kathy uncorked a perfect spiral pass right into his arms. He caught the ball and stood there stunned. And so did the rest of our fellow Marine Corps War College students. And that's how we met our new neighborhood friends, (Lt. Col.) Rusty and Carol Jones and their two sons Rich and Jeff.

One of the big benefits to our boys was that Rich and Jeff Jones, at eleven and thirteen, were just enough older to make them good mentors and sitters for our boys. Matthew and Brendan, at ages (almost) ten and (almost) seven, came to adore these two boys. Their subsequent love of skateboarding can be laid at their doorstep of the Jones brothers.

The academic year was extraordinary. It was done in ten-week trimesters. The first trimester was Policy and Strategy, next came Naval Operations, and the third was Defense Economics. In addition to each of the primary courses, we had to take an elective each trimester. My first elective, strongly recommended by the Marine Colonel who was our liaison officer, was *Foundations of Moral Obligation*, an ethics and philosophy course. My second trimester elective was *Terrorism*, and the third was *The American Military Officer in Literature.*

The Naval War College at that time was not certified to grant Master's degrees. In order to qualify for a masters, we also had to take additional courses at night each trimester at Salve Regina University in Newport. Salve had a partnership with the War College that enabled us to get the masters in one year. I understand that now the War College is certified to award the full Master's degree. Fortunately I had educational benefits remaining on my GI Bill that paid for the Salve portion of the program. The GI Bill had paid for my last two years of undergraduate bachelor's degree work at Mount St. Mary's (thank you FDR), and now there was enough accumulated benefit remaining that I could complete a master's degree program. That was one great deal.

It was quite an academic load with nightly reading and paper writing requirements that were staggering.

Policy and Strategy was a most interesting trimester. We studied warfare from a historic perspective starting with the Peloponnesian War (the Greeks: Athens versus Sparta c: 435 BC) and worked our way forward through the Napoleonic wars, the late 1800s in Europe under Bismark's influence, the First World War, World War II, Korea, and Vietnam, to present day 1984. We studied the high-level policy planning as was developed by heads of state. Then we analyzed the strategic plans supporting that policy that had been developed by the military.

As we worked our way to modern times we developed an understanding of the decision making processes made by presidents of the US and earlier heads of state based on information provided by the diplomats, intelligence services, and the top generals and admirals of the armed services.

We basically were studying policy and strategy development from the 50,000 foot altitude of government.

Three days per week we met in small seminar groups (10 officers each) facilitated by a civilian PhD and either a Navy Captain or a Marine or Army Colonel. I can't recall the name of the Navy Captain who co-facilitated our discussions, but I do fondly remember our civilian PhD. His name was George Baer and he was a brilliant, open-minded person who seemed refreshed by my more liberal perspectives. George and his wife became good friends and we exchanged Christmas cards for years before he passed away.

The seminar discussions were lively and free flowing. All points of view were encouraged. As a more liberal thinking person coming from a civilian world perspective, I was often at odds with my more conservative active duty counterparts. But my brother officers only chided me for my liberal outlook. Back then the military, particularly the senior officers, in general, had been conservative leaning. And these were the Reagan Administration years. Liberal thinking was not a common occurrence in the military. Nor, I was later to discover in a subsequent assignment, was it encouraged or tolerated.

On the days when we did not have our seminars, guest speakers would address the entire student body in the main auditorium called Spruance (named for famed WWII Navy Admiral Spruance, the brilliant and modest fleet commander in the Pacific). Guest speaker viewpoints also added fodder for our seminar discussions. The War College Junior Class consisted of Navy, Marine

Corps, Army, Air Force, and National Guard officers, Navy Lieutenant Commanders or Army, Air Force, or Marine Majors.

We also had a "Senior" class" that consisted of officers from all of the armed services and their Navy equivalent Commanders and Captains. Scattered in were CIA and NSA (National Security Administration) students. Those guys kept a low profile as might be expected.

The Naval War College also hosted two international classes: a grouping of junior mid-career officers, and senior mid-career officers in programs that paralleled the basic Naval War College course, but without access to some of the more sensitive classified information. These officers were from England, France, Belgium, Germany, the Netherlands, Austria – virtually all of NATO – as well as Morocco, India, Australia, and a range of other allied countries around the world.

The first elective I had chosen, on the advice of the Marine Liaison Officer, Marine Colonel Larry Gaboury, was the philosophy course called Foundations of Moral Obligation. It was the crown jewel of the academic year. The course was originally designed and implemented in the mid-1970s by Admiral James Stockdale.

Admiral Stockdale, while in captivity in the Hanoi Hilton, the horrible North Vietnamese prison, came to understand that mid-career military officers, while long in technical acumen, were short on philosophical foundations. Stockdale had immersed himself in philosophy courses when he was in graduate school on detached academic duty from the Navy. That gave him the solid basis to withstand the rigors of POW life. But he saw that too many of his colleagues lacked the philosophical strength to withstand the beatings, torture, and psychological manipulation.

We were most fortunate that Admiral Stockdale happened to be in Newport on the first day of our Foundations of Moral Obligation class. He became our guest lecturer.

The ninety-minutes of that class were astounding. We were in a double sized classroom thanks to the middle partition being retracted. In addition to the 35 registered classmates, most of the wives and a number of others had crowded the room to listen to this man and his remarkable story. Those auditing the class stood along the walls of the double classroom. We who had enrolled in the course were at our student desks which formed a ring around this forum.

Stockdale started pacing up and down the open area between our desks and began unfolding his story.

He had been the ranking Navy officer POW held in the infamous Hanoi Hilton prison in North Vietnam for over seven years. As a Lieutenant Commander, and squadron commander, he was shot down over North Vietnam in his A-6 Skyhawk. He successfully ejected, but one of his legs was broken on his landing.

Local North Vietnamese farmers and soldiers soon surrounded him, and he was brought into military custody.

His first four years were in solitary confinement and were broken up by regular torture sessions. His leg was re-broken several times because he refused to provide any information to his captors. But because of his strong background in philosophy, particularly the Stoic Philosophers, he had learned, one, that he could resist torture up to a certain point, and, two, that everyone has a breaking point and must accept that fact.

As he conducted his forty-five minute lecture to us, it was evident that Stockdale limped badly. As he limped up and down addressing our class, we saw that his left leg was turned inward at a 90 degree angle. For three-quarters of an hour, Admiral Stockdale mesmerized us as he limped up and down the empty space between our desks and told the story of his captivity.

What was so powerful was the fact that he had been separated from everything he knew as a civilized western man. He had been tortured, his leg re-broken repeatedly and badly re-set. Following one torture session early in his captivity that left him unconscious, he learned the code he could use to communicate with his fellow POWs. He had been left unconscious on the floor of his torture room. His tormentors apparently stepped out for tea. When he regained consciousness, he looked up and saw that he was partially lying under a table. On the underside of the table some earlier detainee had carved an alpha-numeric grid minus the letter "K." This is what he saw:

	1	2	3	4	5
1	A	F	L	Q	V
2	B	G	M	R	W
3	C	H	N	S	X
4	D	I	O	T	Y
5	E	J	P	U	Z

He immediately realized the value of this discovery. This became the unbreakable tap code that Stockdale imparted to his fellow POWs. Lord knows who first put the code there, but the North Vietnamese never could break it.

He showed us how the tap code became unbreakable, and how it sustained his fellow inmates. He issued directives, via that "Tap Code" stating that even though his brother POWs would be broken under torture, he mandated that they were to continue to resist coercion by their captors.

It all started with scratching on a wall. If Stockdale heard a door opening and sensed that someone had been deposited in one of the cells on either side of his, he would begin the patient process of communication. First he would tap "dat-da-da-dat-dat (shave-and-a-haircut). That is a peculiarly American signal. Most Americans, probably 99.9 percent of them, would identify that tap signal and respond: dat—dat, for "two bits". Then he would tap or scrape another twenty-six times on the wall. And he would wait. If he got no response, he would scrape the wall twice and repeat shave-and-a-hair-cut followed by the twenty-six taps or scrapes. At some point, the newly arrived American prisoner would recognize that the twenty-six taps represented the American alphabet. Then he would finally respond: tap-tap.

Stockdale, at that point would slowly explain in plain English, spelling out in the conventional alphabet, what the tap code was. It might take days or weeks, but ultimately the guy on the other side of the wall would get it. From that point, communications proceeded rapidly.

It finally got to the point, when there were too many POWs to keep in isolation and they had to be grouped in cells, that the communications could go completely unimpeded. Sometimes the code was tapped on the wall. Sometimes it was scraped. Sometimes later on when there were more officers captured, it would be coughed across the compound, or it might be 'swept' across the compound by POWs assigned to sweep leaves and other debris away from the open areas. Or it might be finger flashed below the doorways so other prisoners could 'see' the message without alerting the guards.

They also could rotate the code on a daily or weekly basis, for example making B the number one and A the number twenty-six in the code.

He got the hundreds of POWs to give an inventory of their talents, and then set up training classes. By tap-code, the POWs learned music theory, English Literature, advanced mathematics, astronomy, and more.

The mental acuity gained by him and his fellow POWs was staggering. Years later I went to a lecture in Rhode Island where one of his fellow POWs was the guest speaker. At one point in his talk, he simultaneously tapped out what he was saying. Spooky.

Stockdale's leadership message was clear and unmistakable. If you were tortured, you would ultimately break. Resist torture as long as you can, and then break. But what would happen was that when you were dragged back to your cell, you would receive messages of love and encouragement from your fellow prisoners. The next day you would be brought out to sign a confession. At that point you were to refuse and subject yourself to more torture.

The overwhelming percentage of POWs bought into Stockdale's message, and their resistance baffled their North Vietnamese captors.

For his extraordinary leadership under the most adverse of conditions, Admiral James Bond Stockdale was awarded the Medal of Honor, our nation's highest award for bravery.

He was retired and living in California when we got to the War College, and his hand-picked successor, civilian Professor Joe Brennan, had been teaching the course by himself for about seven years. We were fortunate that Admiral Stockdale happened to be in town the day of our first class and he came to co-teach it that first day with Joe Brennan.

I remember being stunned (as were my classmates) half-way through his lecture when he said, "Really, I didn't have it so bad. My wife had it worse." We were sitting there trying to parse that idea when he went on to explain that his wife, Sybil, suffered severe mental and emotional pain, plus two nervous breakdowns, the direct result of dealing with the United States government. A government that initially refused to even admit that there were American POWs, and that had forbidden the POW wives to communicate with each other or the press.

But she prevailed, and in the end had become such a pain in the neck to the Nixon administration that she was called once a week and briefed on the latest POW negotiations by the Secretary of State, Henry Kissinger, himself. She ended up leading a delegation of POW wives to the Paris peace talks. That move had the effect of creating global awareness of the plight of the POWs and facilitated their release. One tough woman. Kathy and I had the pleasure of meeting her in Newport.

We took a break from Stockdale's presentation, and when we got back, spent the next forty-five minutes asking him questions. It was amazing.

The remaining nine weeks of that class were equally amazing. Joe Brennan, a philosophy professor *emeritus* from Barnard College, was a brilliant teacher. Each week he took a different theme – moral courage; what is worth resigning for?; our biggest moral challenge; can you share intimate personal information with your military colleagues?; etc. – and spent the first half of each class weaving a theme that might range from the Book of Job to Wittgenstein to Camus to the lyrics of Bob Dylan. It was dizzying. The second half of each week was the discussion. It was some pretty heady stuff, and for most of these mid-career officers, it was the first philosophy course they ever had, and the first opportunity they ever had to conduct an introspective examination. All of us loved the course, even the foreign officers.

Admiral Stockdale instituted the course because he had discovered that, in captivity, some officers survived the ordeal admirably, and others intellectually and emotionally crashed and burned. The difference was the moral foundation each one had. Many, but not all, with a strong religious upbringing tended to be survivors. Others who were good survivors were not necessarily religious but had a solid value system and moral foundation in the home as they grew up. Stockdale admitted that he fell in the latter category. However, he had taken some serious philosophy courses when he was attending graduate school at USC while in the Navy, and it was the great philosophical leaders in history who got him through his POW experience.

The captured officers that crashed and burned – who ended up collaborating with the enemy – had weak moral foundations. They, fortunately for America, were of small numbers. When Stockdale got back to the USA, he tried to bring charges of treason against these POWs who ended up doing propaganda interviews and broadcasts for North Viet Nam. His accusations were completely buried as those in power did not want to admit that such behavior could have happened.

Each week we submitted an eight-page paper on that week's theme, and by the following week we had at least one full page of Joe's critique of our papers typed in single space on his manual typewriter. His comments were always extensive and I wondered how he could do that for thirty-five students each week. He was over eighty-years old when he taught the course.

About two-thirds of the way through the course I had an opportunity to chat with him as we walked down the hallway. We had just been assigned the theme "The Biggest Moral Challenge You Have Faced as an Officer." I was telling him that I had a problem writing my paper because (1) I had never served in combat, and (2) I was a reserve officer with far less active duty than my fellow classmates, and I had never really had to deal with a moral issue as an officer. Joe (as he wanted everyone to call him) said that I should write about something in my civilian life. That was easy for me after my experience with the senior managers in the Claims Division at Pennsylvania Blue Shield. Christ, was that ever a moral quagmire.

I asked Joe if he got a lot of papers about the moral challenges officers faced in war. His reply surprised me. He said that when he had first given that assignment he expected that most of the papers would deal with the moral challenges of combat. What astounded him was that these mid-career officers had faced their worst moral challenges when, after their combat assignments, they were assigned either to their service headquarters or to the Pentagon.

It was always an issue of the admiral or general demanding that the junior officers do what they commanded rather than do that which was morally correct. This had to do more with weapons procurement programs and the like that had started years before. If over time these programs no longer had usefulness, sometimes the top brass wouldn't bring the program to a halt. Therefore, when a mid-career officer was assigned to work on one of these outdated programs, they were shot down by their bosses when they tried to show that the program had no usefulness.

That just blew me away.

* * *

The second trimester was Naval Operations, and I won't bother you, dear reader, with an accounting of those ten weeks other than to say I collaborated with a Navy Lieutenant Commander and an Army Major on a brilliant paper dissecting the failed hostage rescue attempt in Iran. It was a dandy of a paper and got us a grade of 'A.' The rest of the course sucked big time. Within a year, the Naval war College had replaced it with a broader course called Military Operations, which dealt with all of the services in a Joint perspective, rather than the Naval service-centric NavOps that focused on deep water operations of aircraft carriers, a subject of little interest to Air Force, Army, or Marine officers.

The elective that second trimester was Terrorism and I did a well-received major paper on Carlos the Jackal, the terrorist who masterminded the

Munich massacre in 1972 and the kidnapping of several OPEC ministers a few years later.

Third trimester was Defense Economics and Decision Making and was all about how the military complex works. We talked about Force Planning, Budgeting, Procurement, and dozens of things that would narcotize the average person. But if you were going to do a tour at your service headquarters or the Pentagon, you had better know this stuff. As famed Marine General Lew Walt once said to a young Captain who complained about having to do a tour at Headquarters Marine Corps, "If you want to be an Arab, you have to go to Mecca." Somewhere in my personal archives is a paper I delivered for this class. The two instructors, one a Marine Colonel, and the other an Army Colonel, said we could do anything – a paper, a song, a poem if we wished – as long as it illustrated that we had a basic understanding of the Force Planning process.

I chose a to do a poem with illustrations drawn by me. It was called "The Flat Footed Floogies,"* and it was as close to a Dr. Seuss poem as I could make it. It got rave reviews and an 'A.' (I think the illustrations are still somewhere in my Marine Corps 'box.'

The elective I chose for the third trimester was designed and delivered by one of my fellow students. US Army Major Rick Kerwin had just completed three years of teaching English Literature at West Point before lucking into a year at the Naval War College. He was impressed, as we all were, when the great American author Tom Wolfe came to speak at one of the lecture series events in the fall. Wolfe had written *The Right Stuff*, the story of the development of our Mercury astronaut program. His novel was a current "Best Seller." The focus of his talk was about how in the majority of military literature American military officers were often portrayed negatively.

Inspired by Mr. Wolfe's lecture, Rick designed and got permission to deliver an elective curriculum in our third trimester that covered major works of literature from *The Red Badge of Courage* through World War I poetry, and books like *All Quiet on the Western Front*, *Catch 22*, and many other works. It was a lively class.

My three major courses at Salve Regina University at night were The Law of the Sea, Latin America, and something else I can't remember.

The reading and writing requirements were non-ending, yet we had a wonderful social year that included some formal gatherings, (the Marine Corps

Birthday and the annual War College formal ball at Marble House in Newport), as well as many informal luncheons, picnics, and other casual events.

It was a great academic year, we made some good friends, and it was over too soon.

*THE FLAT FOOTED FLOOGIES

When I came here to go to school, a Colonel said to me,
B.J. keep your eyelids up and see what you can see.

Through Policy and Strategy I kept careful track,
Though in Nav Ops I just didn't have the knack,
But in Force Planning a story was heard to be told,
I'll relate it here now, though you may think I'm bold:

The Flat Footed Floogies were planning their forces,
All that they needed were ninety-nine horses,
Then they would have parity once and at last,
With the Pointy Eared Dingbats who haunted their past.

The Pointy Eared Dingbats, with complete lack of morals,
Never took time to rest on their laurels;
They bought and they bred, they expanded their forces,
By buying and grooming hundreds of horses.

The Floogies you know had horses indeed,
Fifty descended from one really great steed;
With power and style, though not very tall,
But getting old now, nearly ready to fall.

So the problem was given to young Floogie-type hirelings:
Figure a way to augment these sirelings.

Action Officer Floogies, a small little bunch,
Talked over the problem at a Pentagon lunch;
National Interests and Objectives topped off their list,
Of things to consider in assessing the risk.

Economic constraints just can't be ignored,
When lobbying Congressmen so easily bored;

But Political matters badly finessed,
Could bury them all in a terrible mess.

Security Issues are salable morsels,
When procuring brand new speedy young horsels.

But though National Strategy has Resource Constraints,
Technological Advances can speed up Old Paint,
To make him run faster and climb and turn quickly,
And outfight the Dingbats and make them look sickly.

"We'll need some more staffers," a young Floogie said,
"Two hundred at least will bring to a head,
This problem of planning a force strong enough,
To counter the Threat from Dingbats so tough."

Another young Flat Foot, advanced for his years,
Said, "Two hundred's too little to turn all these gears,
Like Threat and the Congress, and our Allies' concerns,
Jack up the numbers and throw in some interns."

"Six hundred should do the job really quite neatly."
And so 'twas decided to fund them completely.
They proved that Existing Forces just weren't good enough,
Available horses needed more stuff.

Then technology increased the cost of the beasties,
From one hundred bucks to ten thousand eachies.

But that didn't stop our brave little band,
From plodding on forward, Assessment in hand;
Of Deficiencies, Risk, Alternatives, Forces,
And Fiscal Guidance, just to buy horses.

To satisfy Allies, and deter Dingbats,
Horses were needed that surely could slingback,
The arrows and lances and weapons of bad guys,
Who threatened security of all of our Allies.

Research money was granted, consultants brought in,
And the sound of them all, made a terrible din.

A Chance to Serve

Beltway bandits called contractors bellied up to the feast,
And agreed to produce a remarkable beast.
Kevlar horseshoes, Teflon bridles were added to steeds,
In attempts to enhance capabilities of breeds.

Analysts recommended passive sonar and such,
Surely that cannot be considered too much;
Then Sidewinders and cannon and rockets galore,
Enhanced the old product, but wait, there is more.

Electronic additions, infrared missiles,
Mavericks, AMRAMS, TACFIRE and whistles;
And nuclear warheads mounted on flanks,
Surely would decimate enemy ranks.

With all of this gear we will multiply forces,
Get ten times the value on ninety-nine horses.

"One thought," said a Floogie long considered a nerd,
O&M* must be budgeted for this equestrian herd;
Technicians with shovels and acres of hay,
Will ensure our success in battle today."

His colleagues all cheered, they surely were awestruck,
They came back from Congress with billions of bucks.

Planning went on, but soon they did find,
From a critic who only meant to be kind;
That while they intended a force of great harm,
They'd burdened a horse that couldn't leave his barn.

And so, my dear liebchen, this tale now will finish;
But keep your chins up, let your hope not diminish;
For in Pentagon halls you surely will find,
Some beasts of a more unusual kind.

At the War College we learn about Force Planning in classes,
While at the Pentagon you'll find some real horses' asses.

* Operations and Maintenance

It's a wonder that I wasn't expelled on the spot. But the interesting part of it was I got an 'A' for creativity from the Marine colonel who ran the class.

I remember one class in that trimester when we were discussing the ethics of following the orders of senior officers at either our service headquarters or at the Pentagon. The question raised by the Army colonel instructor dealt with how we, as subordinate 'Action Officers' at our service headquarters or at the Pentagon would respond to an order that went counter to our moral training.

One of my classmate, an Army Major and West Point graduate, said that moral decisions were the province of General/Flag officers and it was their problem. We should only follow orders.

Needless to say I was out of my seat in an instant and said, "Whatever happened to Duty, Honor, and Country? Have you forgotten your oath?" Fortunately the instructor stepped in before things got out of hand, but it was a revealing moment.

At what point in an officer's career does he or she say, "I won't do that because it is immoral (or illegal or fattening)?"

17.

S-3, Twenty-Fifth Marine Regiment Worcester, Massachusetts

1985-1987

The transition from the one-year academic experience at the War College was challenging. As we finished our military graduate school experience, I was reluctant to let go. I was offered an extension of Active Duty for 60-days to stay on as a Global War Game controller for the annual game played at the Naval War College. I readily grabbed at that offer and took that assignment, which extended from our graduation in June through early August. Kathy had already landed a job in Providence, and we had looked for and settled on a nice little ranch house in Barrington, RI, where we could resume the effort of raising our children in a stable environment. Now I needed to return to my prior Reserve Weekend Warrior status.

Knowing that this was inevitable, in the spring before graduation from the Naval War College I had applied for the job of S-3 (Operations Officer) for the 25th Marine Regiment headquartered in Worcester, MA. With a solid recommendation from one of my instructors at the Naval War College, Col. Stan Pratt, I was awarded the job. It was a prestigious position being that there were only three infantry regiments in the Marine Corps Reserve, and another nine of them in the Marine Corps as a whole. I was pretty proud of myself to be in such elite company but shouldn't have been.

The assignment turned out to be more challenging than I thought it would be. It seemed that there were several other officers from within the regimental staff that had applied for the job. Most had been with the regiment for years. And here I was, a newcomer, an outsider, taking a position that several of them perceived as an entitlement that was rightly theirs.

It was not an easy two years.

Before I go further, let me explain a little about what had become a restrictive practice in the Selected Marine Corps Reserve, and a practice that should be eliminated. On active duty, an officer (or enlisted man) can expect to change duty stations every three years or less. That was the nature of being on Active Duty. For example, you might be an infantry officer at Camp Lejeune, NC, and after three years you might be assigned to a three-year tour as a recruiting officer in Buffalo, NY.

In the Reserves, however, 'Homesteading' was a common practice. If you were an officer, it was fairly easy to find a billet on a battalion or regimental staff and rotate from billet to billet within that organization for most of your career.

To complete a successful twenty- (or even thirty-) year career as a Marine Corps Reserve Officer, one needed to follow certain 'yellow brick road' footprints. The steps began when you completed serving three years on Active Duty. In the Vietnam era, that usually meant that you departed Active Duty as a senior First Lieutenant. If you were lucky enough to land a position on a Battalion staff, you basically had it made if you were careful.

As a senior First Lieutenant, your first tour would be at the Company level as a platoon commander, or it might be staffing one of the battalion shops (S-1 Administration, S-2 Intelligence), or being an assistant in S-3 Operations or S-4 Supply/logistics). Within a year you are promoted to Captain and eligible to be in charge of at least Admin (S-1) and Intelligence (S-2). Now three more years have passed. Time to get a new job. Bingo! Low and behold an opening has occurred in the Operations (S-3) section of the battalion staff. You grab the opportunity and spend another three years in the battalion as an assistant in the operations shop. Then your tour is about to expire and you apply for a company command, and ("Surprise! Surprise!" as Gomer Pyle would say), you get selected to be company commander of one of the companies in your battalion.

Now you are finishing your tour as Company CO and approaching promotion to Major. The savvy officer (I was not one of them), rotated back into the battalion he had left only three years before. Now you were elegible for one of the S- shops as a senior Captain; three years after that (now as a Major) you become the Assistant S-3 (Operations) Officer; and then maybe you squeeze in another three years in another of the shops.

Now comes the bigger prizes. You have at least three years of Active Duty USMC Commissioned service and have served in the Marine Corps

Reserve component for eighteen or twenty more years. To your credit, you have also commanded at the Company level. Now you've been selected for, and maybe already promoted to, Lieutenant Colonel, and the battalion CO or Regimental S-3 (Operations) job comes open. In most of the battalions and regiments in the USMCR, if one is a betting man, the percentages favor the in-house (homesteaded) candidate getting the job. Next thing you know is that if you want to earn promotion to full colonel, you need a two-year tour as Commanding Officer of a Battalion, or, more specifically, the Battalion where you have homesteaded for most of your career.

Outsiders and Irish need not apply.

Forget the part about the Irish. It was a joke referencing the time in our history (1847 through the early 1900s) when being an Irish Catholic automatically disqualified you for full-time employment in many major industries.

I only wanted to point out the discriminatory similarities that existed within the Marine Corps Reserve billeting assignment system.

I was not fully aware of the politics of all this when I applied for the Regimental S-3 position. That was mistake number one. Chalk it up to my naivete. The next mistake was that I had had only brief experience on a battalion staff, and that was when I did a little more than a year with 4th Engineer Battalion in Baltimore as the S-1. I should have had more battalion level staff officer experience before I plunged into this new assignment.

And even though I had de facto run Hotel Company when the company leadership failed to step up to the plate, I had not officially 'commanded' an infantry company.

Arriving so unprepared for the regimental S-3 assignment, I really did not have a firm grasp on how an S-3 officer professionally functioned in this environment. I did, however, blindly trust to the fact that my staff would help me on the learning curve, as would the I & I, a tough guy Active Duty Lt. Col.

Wake up Lt. Col. Power and smell the gunpowder. That was not about to happen.

It was a shock to me that when I asked for help in figuring out some of the job requirements, I was met with a lot of silence. The message was pretty clear: you took the job that some of us wanted.

Somehow I was able to function through two years of monthly drills and two-week annual training duties, but I felt very alone. I also realized that I did not have a core of solid NCOs and Staff NCOs who would help me become a better staff officer as had happened in Japan, Camp LeJeune, and Harrisburg. The SNCOs and NCOs that I had on staff gave their allegiance to the long-time homesteaders.

Mercifully it came to an end in July, 1988, when we got back from annual training at Camp LeJeune. The regimental commander was most kind, but he was perfectly clear. He wanted me to step down from my job as S-3 as I seemed to be having problems getting cooperation from the staff. He offered to place me in another job within the regiment. I thanked him but said no.

Interestingly, he gave me a "water-walker" fitness report. He said he knew I was a terrific officer and that I had put my best effort into it, but he also saw what I was up against as an outsider. I really respected that man.

So now it was time to move on.

18.

The Marine MTU at the Naval War College

1987-1989

Welcome to the most eye-opening experience I ever had in the Marine Corps Reserves.

I discovered that there was an MTU (remember that these units were unpaid but earned Reserve Retirement credits for drill participation) in Rhode Island that drilled at the Naval War College. Its mission was to provide Global War Game support. I joined the unit hoping it would be a transitional position while I sought another paid billet.

When I joined, I found the unit Commanding Officer to be a terrific Marine Colonel who had the respect of all the officers in the unit. By the way, I may not have mentioned that most of the MTUs I had been affiliated with were staffed completely with officers. I had joined this unit in September of 1987, and this colonel retired in June of 1988. His replacement was another colonel in the unit. The new CO seemed to be okay, or so I thought.

I am not going to dwell too much on this episode, but it was the crushing moment of my career.

In the summers of 1988 and 1989, our unit was asked to provide additional support to Commanding Generals' Symposia that were held in the Washington, DC area each year.

I was one of the more flexible members of the unit regarding my job (at that time I was an Outplacement Consultant contracting with Boston area consulting groups to provide services to regional companies that were downsizing). I had a lot of flexibility as to when I did consulting and when I took advantage of paid Marine Reserve opportunities. In the summers of 1987 and 1988 off I went with the good (but outgoing CO) and his XO (the heir apparent) to provide support services to convocations of Marine Corps General officers who had been called in from all over the world to 'War Game' some global threat scenarios in Washington, DC.

I decided that I would need wheels when in the DC area because we would be going back and forth from Headquarters Marine Corps, i.e., HQMC (which was then in Arlington, VA), to Quantico, VA,. In each of these two years my two colonels flew in, I picked them up, and we were all billeted at the Sheraton in Arlington almost next door to HQMC.

For the next two weeks I chauffeured the two colonels wherever we needed to be. Then I ended up driving them on personal errands. That part pissed me off. I'm willing to do my part to fight off the Soviet threat, but running personal errands for a Colonel? It was bullshit, but I did it.

Any time you are in a situation like that, you cannot help but make a personal connection with your colleagues, even though they might outrank you. The outgoing CO was content to have genial conversations about non-controversial topics. His heir apparent Colonel liked to talk politics. It was evident that, while he was in civilian life an attorney employed by the Commonwealth of Massachusetts, he hated the liberal democratic governor, Michael Dukakis. That governor ran for president in 1988 against the eventual winner, George H. W. Bush, a Republican.

As we were driving on yet another endless personal errand, the colonel made some disparaging remark about the Democratic contender. I countered with, "Yes, but the Bush campaign was shameful when it used the race card and brought out the fact that a convicted black criminal had been given early release by the Democratic governor and then went on to commit a horrendous crime. I said that it surely wasn't an indication that the Democrat governor deliberately meant to turn a violent criminal loose on society.

I got a frosty silence. But I didn't give it much more thought.

I should have learned from a lot of past experiences to keep my mouth shut. I was to learn just how evil some people can be when they have power over you.

So now we (the MTU) are preparing to support a Global War Game at the Naval War College in the summer of 1989, and the former "heir apparent" is now the new CO. At one of our unit planning meetings early in 1989 we got into a major discussion of just how we could support the War Game players, both civilian and military.

From my experience as a Global War Game Controller going back to 1983 at the Pentagon and again in 1984 at the National War College, as well as supporting the Global War Games at the War College in 1985 and 1986, I remembered a unique tool that was commonly used at the Pentagon to prepare senior officers and Political Appointees when they went to testify before Congress. I had also seen this tool used to brief players in Global War Games both at the Pentagon and at the National War College at Fort McNair in DC.

It was called a 'Smart Book.'

I suggested that, because the 1989 War Game was to have a Latin and South American focus that we should prepare Latin American regional 'Smart Books' for the players. This was met with an enthusiastic response and approved by our new CO, and so the assignments were parceled out to the members of the unit.

As there were at least twenty-five countries represented by island nations as well as mainland countries, I suggested that we (twelve or so, maybe fifteen) officers each take a country or two or three. Major 'A' would have Guatemala, Costa Rica, and Belize. Lt. Col. 'B' would have Columbia, Peru and Bolivia, and so on. Each member of our unit had a responsibility to produce a background political summary, along with appropriate military, demographic, industrial, and agricultural information on one regional country. All this would be assembled into a tabbed three-ring binder for easy reference.

In June, as the War Game approached in July, I polled the unit officers who had accepted assignments to do the background information on each country and discovered that about 10% of the information had been gathered.

I realized that we were now on a short deadline, so I contacted the Marine Reserve Colonel who had been my counterpart student at the Naval War College in 1984-1985, and who now was part of our MTU, and asked him to help me out. He readily volunteered and I assigned him several South American countries to research, and I took on the balance of them myself.

Bottom line was that we produced a working 'Smart Book' for Latin America. It was a three-inch binder with tabbed dividers for each country. The actual Marine Colonel who was in charge of the Southern Command cell for the war game (i.e., Latin America) gave it high marks for a unique way to have major reference material available.

When we finished supporting the war game, I felt that we had managed to accomplish the mission.

In August of 1989, I got a call from a Major in Richmond, VA, who was a Marine recruiter with the specific assignment of finding a Major or Lieutenant Colonel who would take an assignment to a Pentagon RAU working in J-4 (Logistics) in the Reserve and Industrial Mobilization branch. The limitation to such service being that said officer must be able to fly to DC during the work week and work from two- to four-days at a time with his Active Duty counterparts.

It seemed to be a dream job for me and I jumped at it.

So I gave my notice to the MTU, and had a nice exit interview with the CO, and reported for duty at the Pentagon.

Administratively, when one departs from any command, one gets a fitness report with the indicated reason for it being "Change of Command." I didn't give the fitness report a second thought as 1) I thought I had a good relationship with the CO, and 2) my fellow officers gave me very high marks as a leader and many expressed to me later that they thought I was on a fast track for promotion to the rank of Colonel.

All too soon I was to be sorely disappointed with the Marine Corps and its officer evaluation system.

19.

Final Reserve Assignment with the Pentagon RAU

1989-1994

In September of 1989, I reported to the J-4 (Logistics) branch of the Joint Staff at the Pentagon in Washington. Actually the Pentagon is across the Potomac River in Arlington, VA, but everyone refers to the Pentagon as being in Washington.

My new assignment within the J-4 (Logistics) branch was in a section dedicated to Reserve and Industrial Mobilization. My timing for this assignment could not have been better. Within twelve-months, our little department was going to become the "eye of the storm" as our country began preparations for moving a staggering amount of equipment and hundreds of thousands of soldiers, sailors, Marines, and airmen to the Persian Gulf.

The assignment required that I perform my monthly drills not on weekends, but during the work week where I would be fully integrated into operations with my Active Duty counterparts. Once I got the lay of the land and had been given some great advice from the Army Reserve colonel I would be working closely with, I came up with a monthly plan.

The Marine Corps, while requiring my presence at this assignment, would not be paying for travel expenses. Therefore, to save money I booked round trips from Providence to DC that would transport me down one month and return the following month. Simultaneously I booked a round trip from DC to Providence for the same months but planned the trip around my drill dates. With my airline of choice, USAIR (now non-existent), this became the most cost-effective way of traveling. Had I, for example, paid for a round trip beginning Monday and returning Thursday of the same week, the cost was more than $500. By doing it on alternate months, I was only spending around $160 per flight.

For example, I might fly to DC on October 10 and return October 14, and then repeat the round trip in November on the 16th and 20th. But by buying the tickets so that they appeared to be a month apart in travel, I got them from the airline at a greatly discounted rate. If I bought a round trip from October 10 through 14 and another trip from November 16 through 20, I'd be paying so much for airfare that it would gobble up all my Reserve pay which at that time was about $450 a month.

A pain-in-the-ass way to make a living. And I'm reminded of my colleague back at Hotel Company, Lieutenant Jack Kotanchek, when he said in the middle of an outrageous downpour at Jungle Warfare School in Panama, "I'm only in this for the money."

Fortunately another way to save money was to bunk at my in-law's house in Arlington. The food and the company were excellent.

Those first several months in the Pentagon Reserve Augmentation Unit were eye-opening. I got a chance to see how the Pentagon operated. Our shop was not large. We had two Navy Captains who were the equivalent of CO and XO. Then within the shop we had two officers in charge of Industrial Mobilization. When I joined, both were Lieutenant Colonels (one Army, one Air Force), but both were soon promoted to Colonel by their respective services.

In the Reserve Mobilization part of the shop, we had an Army Reserve Colonel, an Active Duty Army Colonel, and a woman Air Force Reserve Lieutenant Colonel. Within a year we also had a couple of sharp young officers, one a Navy Lieutenant and the other an Army Captain, both graduates of their service academies, to assist us in our daily efforts. Supporting our efforts were an outstanding Navy Petty Officer First Class and a terrific civilian secretary.

Save for the two junior officers just mentioned, the rest of the officer corps in the Pentagon from the rank of Major on up were all graduates of one of the services' senior war colleges. This had become a requirement under the Goldwater-Nichols Act of 1986 that mandated a higher level of professionalism and cooperation between the armed services. That act made a Pentagon tour more than just an item checked off on one's career to-do list. It now required a professional cadre of officers and enlisted who were educated to be more concerned with "Jointness" than parochial service interests.

Consequently the environment in the Pentagon became one of professionalism and collegiality. Except for the occasional "dumb Marine" joke, the professionalism and inter-service cooperation among these officers was extraordinary.

There was an additional benefit to my serving in the Pentagon RAU. My shop, Reserve Mobilization, had the authority to give its participating Reservists additional Active Duty days each month. This meant that in addition to my regular four Reserve credit points, I would accumulate an incredible number of additional active duty days that not only added to the monthly paycheck but also built up an exceptional number of additional days' credit toward retirement.

The time I served at the Pentagon with my Active Duty counterparts became the equivalent of an extremely advanced schooling in Joint Operations.

Within one's service (i.e., Marine Corps, Navy, Army, Air Force), a career officer got to know the inner workings of that particular service – its policies, plans, and operational capabilities. But when one served in a Joint Command, and the Pentagon was the epitome of a Joint Command, that individual had to learn how the services coordinated and operated in an environment that was far removed from the old service centric model.

It had become so important to the services that a career officer could not advance in rank unless he or she had been certified as serving in a Joint Command for at least three years.

It was a good policy that promoted a professionalism that served the country well.

My initial time in the J-4 Reserve Mobilization Branch was like attending an incredible school. I learned the practical application of my lessons from the Naval War College. The timing was fortuitous.

I arrived in September 1989 and served an initial apprenticeship through June of 1990. Then in early July the real world intruded. Saddam Hussein, Iraq's 'president for life' (and it wasn't to be a much longer one), had ordered his army to invade Kuwait, Iraq's neighbor to the south that bordered the Arabian Sea.

But before that could be resolved, my own personal career crisis had arrived. I knew that I was in the promotion zone to colonel. The promotion board met at Headquarters Marine Corps in May of 1990. I had been assured by most of my peers that because of my "water-walker" fitness reports submitted by three-star generals heading the various Global War Games, coupled with my selection for and successful completion of the advanced Naval War College program, coupled with my successful (at least on paper) tour as S-3 for the 25th Marine Regiment, I was a shoe-in for selection to the rank of colonel. My peers in my former unit, the MTU in Newport, were in unanimous agreement that I should have been promoted. And so I believed.

Well the board met, the votes were cast, and I was not on the promotion list.

I immediately delved into the possible reasons why. I could only assume that something administratively had failed to be done. After a lot of research, I discovered that all the administrative areas had been covered. And then, just after the board met, I received in the mail a printout of the fitness report written upon my departure from the MTU the prior September seven months earlier.

My former CO had submitted my Change of Command fitness report – seven-months late I discovered – and it had arrived at Headquarters Marine Corps only a week or so before the promotion board met in May of 1990. His fitness report showed me to be an 'Excellent' (but not 'Outstanding') officer. The narrative was a decent write up, but there are nuances to Marine Officers fitness reports that I was to learn about.

There is one section of the report labeled 'Desirability' which was an indication of the degree of desirability with which the reporting senior officer would want to serve in combat with the officer reported on. Anything less than a mark of "Particularly Desire to Have" is a career ender. I got "Would Be Willing to Have."

It was a death sentence. Sgt. Bobby Strawser would have been appalled. Lance Corporal Moore, my muse from Iwakuni would have said, "Just one good fuckin' deal after another, right Lieutenant?"

My former commanding officer had murdered me with kindness on the narrative portion and in all the category markings. That is to say he wrote a good but bland fitness report that didn't say one negative thing, but that gave no overt praise either. Then in the one category that doomed me, he marked me down. He gave me a decent fitness report, but then buried a bayonet in my ribcage with that single mark.

And he could do that legally.

I had never, up to that time in my career, seen such a glaring example of moral corruption and cowardice. If you don't like an officer, then tell him to his face, and write him a bad fitness report. Within the system, if anyone receives an 'Unsatisfactory' fitness report, he or she has the legal right to attach a rebuttal statement. Once that is done, the promotion board can see both sides of the issue.

I never had the opportunity to rebut anything because he deviously and in a most cowardly manner ruined my promotion chances with this sham of a fitness report. Interestingly, there is a procedural issue here as well. All fitness reports must pass though the hands of, and be thoroughly screened by, the unit's administrative officer.

I knew this officer well, and coincidentally he now happened to be mobilized and assigned to the Marine Public Affairs Unit at the Pentagon.

I sought him out and explained what had happened and he was stunned. He assured me that the fitness report the colonel had written on me was not the one I saw. Apparently, as required, the colonel had shown the administrative officer one fitness report. Then, after the administrative officer had been transferred to another unit, the colonel had replaced the fitness report with the new 'Down' marking. Furthermore, he delayed submitting the fitness report for more than six-months until just before the promotion board met, leaving me no opportunity to rebut it.

I now found myself in administrative Hell.

The first thing I did was call my former CO and ask why he had given me such a fitness report. His reply was that he didn't feel that the project I had completed in support of the Global War Game was very good. He also said that his civilian counterpart at the Naval War College concurred. I told him that his reasoning was no justification for the markings he gave me. He said that he would not change the report.

I called the senior civilian coordinator at the Global War Game and he said that he thought my project was quite good.

I called the reserve colonel who had been in the Senior Class at the Naval War College with me and also with me in the unit. It was he who had given me assistance in completing the Smart Book project. I asked him what he thought of my performance and he thought I was an outstanding officer and a real asset to the unit. He said that I deserved to be promoted to colonel. I asked him if he would write a letter on my behalf to the Board of Correction of Naval Records (BCNR). There was silence on the line for a moment and then he told me he didn't think he could do that.

I knew that he was a very politically oriented career officer who believed that he was under consideration for promotion to Brigadier General. He was afraid to stick his nose out. Another example of moral cowardice. And this was a combat officer in Vietnam! (BTW, he never made general.)

So much for Marine Officers taking care of their own.

I went to a Marine Colonel I knew at the Pentagon and asked his advice. He said the first thing to do was to see if any other senior officer in that old unit would write a letter of support for me. I told him where that got me. He then said that I would, on my own, have to put my case before the BCNR (the Bureau of Correction of Naval Records). I told him I didn't know how to go about doing that and he suggested I get some mentoring from one of the Marine generals at the Pentagon. He gave me a name and a phone number and I called for an appointment.

I met with this two-star (Major General) in his 'E' Ring office (outer ring, first floor of the Pentagon – reserved for the really powerful). I regret that I cannot recall his name. Of all the offices I had entered in the Pentagon, and this included the Secretary of Defense's office, this one exhibited the most power. Of all the other Flag Officer (i.e., generals and admirals) offices I had ever gone into, as well as the various deputy secretary (and SecDef) offices, and there were a number of them, this one blew me away. All the other "Flag" officer and senior civilian offices were cluttered with souvenirs from overseas assignments, framed acclamations, framed proclamations, and photo ops with "Important" people accumulated over a thirty-year career with many overseas assignments.

This general's office was the most austere I had ever seen. At the Pentagon, on the 'E' Ring, all these Flag Officers rated huge offices. I would bet that most of these offices were in the 400 square foot category and cluttered with plaques, framed awards, loot from international tours, and pictures with the glitterati of the international military and civilian community.

Not this Marine general. He was relatively short for a man of such achievement (maybe five foot seven), and modest, a characteristic not often seen in Flag officers, but when it was evident, you knew you were in the presence of someone who deserved his or her rank.

Stepping into his office was like stepping into a monk's cell, but a lot larger. There was not one picture or plaque on any wall. No loot. No pictures. No framed awards. No statuary. No 'I been there and done that' display of ego. No 'Look at who I was photographed with' pictures. There was just his desk, three walls painted white, a panoramic window view over the Potomac River, and little else. Centered on the back wall of the room was his desk, a solid antique-looking wooden item, but of little note. In another corner there was a small couch with a corner end-table, a coffee table, and an upholstered straight-back chair. On the corner end-table there was a small 5 X 7 framed picture of his daughter in her dress whites being commissioned as a Marine officer by her father.

That was the only personal touch in the room.

I felt about two-inches tall as I walked in and absorbed the atmosphere. I had never been in an office so powerful.

Despite the ambience of the room, the General was very warm, gracious, and empathetic. After guiding me to the corner 'conversation pit' he explained the process of appeal and helped me with some other administrative procedures regarding the process.

I went forward armed with little more than moral outrage, but I had a good story to tell. I made my case in writing to the BCNR, and my request for reconsideration was denied. I asked for a hearing and it was granted. I pitched my case before a team of Navy and Marine officers, but to no avail.

The whole damned thing broke my heart. The system was truly unfair. I was told a few years later, by an officer who had served on promotion boards, that on the first day when the boards met, they scanned the hundreds of eligible candidates' most recent fitness reports looking for 'Down' markings on certain critical areas. If one 'Down' marking was found, that candidate's history was set aside and placed in the 'No" pile.

There was no one saying, "Hey, let's look at this officer's whole career and get some perspective on his or her candidacy." At the Colonel selection level, the pyramid really narrowed, and unless the board saw nothing but 'water walker,' they rejected the candidate out of hand.

Now let me grant that the major deciding factor for selection for promotion to Colonel in the Marine Corps and its Reserve component was 'Command.' Okay, so I never formally had command. But other officers without Command had been promoted to Colonel. I only wished that I could have had a fair evaluation. If all the other plusses in my career didn't add up to promotion, then so be it. I can accept a decision based on merit. But I find it hard to accept a decision based on vindictiveness because of my political leanings.

And so it went.

I was crushed. I was forty-four years old and had put an exceptional amount of emotional and physical investment into an organization that had, for a lifetime for me, represented everything I wanted to be. The system had failed me. I now had to make some choices.

I decided it was time now to put my emotions aside and settle in to do a good job at Reserve Mobilization Branch. Maybe next year the Promotion Board would reconsider. We had a war coming on. I pinned my hopes on the following year when it was time for my next annual fitness report. My Pentagon bosses did not disappoint me. They gave me another 'water-walker' report. Those folks were terrific. They had seen what I could do in peacetime, and they had seen what I could do as a staff officer in wartime. They had no hesitation in rewarding me with superlative and 'Outstanding' fitness reports. I'm glad and very proud that I spent the next three years finishing my career with them.

But about that war …

The US and the United Nations were outraged at Iraq's unprovoked attack on its non-aggressive neighbor. President George H. W. Bush began the process of pulling together an unprecedented coalition of nations. Our NATO allies in Europe were behind us, but through creative diplomacy, the president had also brought in an unprecedented coalition that included not only NATO (North Atlantic Treaty Organization – mainly European allies since the end of WWII), but other Arab nations who would not normally consider attacking a brother Muslim nation to join in repelling this incursion.

In early July of 1990, Army Col. Glen Cassidy, my supervisor and mentor at J-4 Logistics, Mobilization Branch, called me and asked what my availability for recall to Active Duty might be. I told him that I was ready to come on Active Duty any time he wanted. He told me to stand by for further orders.

Initially, in that mid-year of 1990, in addition to my two-days of Reserve duty each month, I started being authorized to pick up an additional two to four Active Duty days each month. My reserve pay had bumped up quite a bit as a result, and now I was flying from Rhode Island down to DC on an early Monday morning flight and staying four days (instead of two) until I flew home Thursday evening. I did that at least once a month. It paid off in our monthly financial situation, and, with the extra Active Duty days, in bumping my reserve Retirement pay.

I mentioned earlier that the Goldwater-Nichols Act of 1986 gave new direction and instructions about the US military operating in a Joint manner.

Well, buried in the language of the Bill were also procedures authorizing the mobilization of Reserve forces and Industrial capabilities of the United States under the provisions of a revised Title X US Code.

That doesn't mean anything to you, nor should it. But it was the relatively newly revised, and previously untested, legal means by which our country could mobilize its Reserve Forces and Industrial capability in the event of a national emergency declared by the President.

That legality put Reserve Mobilization Branch, Joint Staff, Pentagon, at the eye of the storm.

Early in July of 1990, my boss, Col. Cassidy, found himself sitting on the living room floor in the home of the Chairman of the Joint Chiefs of Staff (at that time Army four-star General Colin Powell) and preparing documentation that would enable the President of the United States to issue proper legal orders mobilizing the nation's Reserve Forces.

Once that final order was signed by the President, hundreds of thousands of Reservists like me would be legally mobilized. In 1990, from August through December, my monthly Reserve duties (along with a generous amount of Active Duty days thrown in) put me, along with Col. Cassidy, in the middle of the eye of the Reserve Mobilization storm. The Mobilization effort of which he had been a chief architect was now given the code name 'Operation Desert Shield.'

On every monthly trip to the Pentagon, I was assisting in drafting up various documents pertaining to the mobilization of hundreds of thousands of Reservists to support Desert Shield. The air in the building was almost electric with excitement.

In the summer of 1990, it was clearly evident (at least to us) that all of the Military Reservists assigned to billets in the Pentagon would be mobilized. Toward that end, each of our branch heads had requested advanced level security clearances for all of us as we would be exposed to highly classified information we previously had not had access to. I remember filling out the lengthy form requesting an upgrade from my previous Top Security clearance to a more advanced 'Top Secret – Compartmented Information' level. The FBI wanted to know, basically, everyone I had ever met. They also wanted a list of every neighbor and friend I ever had. To their credit, they subsequently interviewed everyone named to see if I was patriotic enough to work at the Pentagon where I had already been working.

While that process was in the works, one day, I believe it was in September, Col. Cassidy called out to the office. He said, "Hey, look at this." We all joined him at his computer and he started slowly scrolling through various operational orders. It seemed that on relatively short notice, and without fanfare, the Marine Corps had deployed two active duty infantry divisions, along with artillery, tank, engineering, and aviation assets to the Persian Gulf; and the Army was mounting out a massive deployment of its forces.

By December all the Active Duty Forces were in place in Saudi Arabia, poised to mount the attack on Iraq. And our small Mobilization Branch was a huge part of that.

Right behind the Active forces, Reserve forces began flooding into the Persian Gulf as well.

The Reserve forces from each of the Armed Services were being deployed to augment their active force counterparts, but first, upon their mobilization, many of them had to undergo short-course crash training on newer weapons systems.

A good example of that was the Marine Reserve tank units. The Army was operating the new and highly sophisticated M-1 Abrams tank. The M-1 had been introduced into the Regular Marine forces in the early 1980s. But the Marine Corps Reserve forces were still training on the old M-48 Shermans, relics of the 1950s.

The M-1 was a remarkable machine. It had a computer capability that allowed it to accurately fire a projectile while driving at full speed over rough terrain. The entire weapons firing capability of this machine was night vision capable. The Marines Reserve tank battalions, having been slated to get these advanced tanks (someday), were now on the fast-track to get trained on the M-1 Abrams. Consequently all the Marine Reserve tank units were initially deployed to Fort Ord, California (the principal Army tank training site), to get orientation training on the newest models.

The logistics of mobilizing all these forces, mounting them out, and getting them to the theater of operations in the most timely manner, started in our shop.

Some of the numbers were staggering. For one particular deployment of just Marine (not including Army and Air Force) Motor Transport assets, the military needed to find the air and sea-lift capabilities to move a fleet of medium sized military trucks whose length, if parked headlight to taillight, would be equivalent to lining them up nose to tail from Boston, MA, to Newport, RI. That is a distance of over sixty-miles! Now we needed ships to put them on.

Add into that all the artillery pieces and related support equipment; all the tanks (at sixty-tons each), HUMVEES, logistics support vehicles, ammunition for every type of weapon, etc., for all the armed services involved in the conflict. The level of logistics support and lift capability for this military effort was staggering, and it all had to be done within a six-month period from July of 1990 to January of 1991.

It was a busy time, to say the least, and it was a time when a bunch of real professionals stepped up to the plate and hit a home run.

By early January, through an exhaustive mobilization effort, we had deployed over 500,000 troops and all their assorted support equipment, weaponry, ammunition, and other logistic support requirements half-way around the globe. We were now poised for the inevitable next step: Operation Desert Storm, that actual part of this humongous bit of theater where there would be shots fired in anger.

An interesting anecdotal episode occurred in our shop in very early January, when, shortly before I was mobilized, I happened to be there for my monthly reserve work. My mobilization orders were in the works but I hadn't received them yet. As I was going about my daily routine, suddenly the door to our shop opened and in walked the J-4 himself – three-star Vice Admiral Plant. He walked right past the office where our CO and XO (two Navy Captains) were in conference and went right to Col. Joe Gould's desk (Industrial Mobilization).

He said, "Joe, what does it take to surge production of Patriot Missiles?"

Joe said, "Two phone calls."

The Admiral said, "Make them."

Over the past couple of months, Iraq's president, Saddam Hussein, had ordered the launching of missiles, known by their acronym 'SCUD', into Israel. The Israelis, with support from US personnel, had been countering those attacks by firing an anti-missile missile called the Patriot (made here in the US). Because of the volume of SCUD missiles fired, the Patriot inventory in Israel was shrinking. General Powell had called Admiral Plant and asked point blank how to fix the problem. So the good admiral did what all good military leaders do when confronted with a question they can't answer. He went to the lowest ranking person who could answer the question.

In the meantime our CO and XO had come flying out of their office and were being as solicitous of the admiral as they could be and asking how they could help. The admiral said, "Don't worry about it. I got the answer."

You might not think this is a remarkable story, but I had seen the Pentagon operating in a routine, pre-crisis mode, and now I was seeing how it operated when it was on a wartime footing. In the pre-crisis mode, the admiral's question would have been bounced all over the bureaucracy before it was answered two weeks later.

What the admiral, and all of the smart admirals and generals were now doing was eliminating the mid-level bureaucrat/career officer place holder and going right to the Action Officer who could answer the question.

When people ask me what the Pentagon was like, I tell them that it was all they feared in peacetime, and all they hoped it would be in wartime.

We brought all of those boys and toys over there, and unless Saddam Hussein had a change of heart, we were going to use them. As the much-respected Chairman of the Joint Chiefs of Staff, General Colin Powell put it on national TV, "We are going to cut Saddam's army off and we're going to kill it."

Any questions?

After everybody I knew in North America had been interviewed by the FBI, my mobilization orders came through in mid-January 1991 and stated that I was to be on Active Duty through at least the end of June. So I had lots of things to do. I had alerted my baseball team earlier in the fall that this was inevitable and left the team in the capable hands of one of my players.

Knowing that Matthew and Brendan (ages 15 and 12) were concerned, and maybe even frightened, about this whole thing, I sat down with them and tried to reassure them that I would be working in an office in the Pentagon and that they had nothing to worry about. But still they were worried. Kathy, being the career military kid, and now a Naval officer herself, took it more in stride. This, I hoped, was going to be no more than a family inconvenience and would be over soon.

While I could have asked for and received paid housing in a local hotel, I opted instead for crashing at my in-laws house in Arlington. Anne and Ed were happy to have me and I had the whole basement, formerly the bedroom for brothers-in-law Tony and Pat, to myself. Plus, Anne was a superb chef, so I knew I would eat well.

Shortly after reporting to my office in the Pentagon, my temporary Top Secret-Compartmented Information (TS-CI) security clearance came through. I had always had a 'Top Secret' clearance, but in order to stand watch in the National Military Command Center (NMCC), I needed a higher level of clearance. The upgrade enabled me to receive the twice-daily briefings from the Central Intelligence Agency as well as the Joint Staff J-3 (Operations) briefing. These briefings were pretty intense and involved a lot of stuff that the general public didn't know, like incredible satellite photography and the actions of Special Operations ground assets (i.e., humans) we had infiltrated into downtown Bagdad.

Representatives from each department of the Joint Staff (J-1 through J-8)[1] rotated on a 12-hour on, 12-hour off watch schedule. From early January, when the air war component of Desert Storm began, and through the intensive, brief and highly successful ground campaign in mid- to late February, through the early execution of the re-deployment (the beginning of bringing troops and equipment back home) stages in mid-March, we all were on a high intensity watch schedule. For five days, I would be on an 1800 (6:00 PM) to 0600 (6:00 AM) schedule. On that shift, when I left the Pentagon after the turnover briefing, I usually drove a couple of miles out Columbia Pike to a classic old silver diner for a bacon and eggs breakfast before returning to Anne and Ed's house and crashing.

I would sleep all day (except on days when Anne would be shouting instructions down to the cleaning service lady who was doing laundry in the room next to mine in Anne's basement), getting to bed around 8:00 AM, and getting up around 4:00 in the afternoon. I would shower, shave, put on a clean uniform, and Anne would give me an early dinner before I had to report for my watch at 1800.

Every five days the watch would shift. For example, I might finish a five-night shift routine, then I would have the next day off and then the day after that have to report for five days on the day shift (0600 to 1800). It was a pain in the neck having to change one's body clock every five days, but that's how it was done.

---

[1] The 'J' sections of the Pentagon are as follows: J-1 Administration and Personnel; J-2 Intelligence; J-3 Operations; J-4 Logistics; J-5 Plans, Policy, and Strategy; J-6 Command, Control, Communications (C3) and IT; J-7 Joint Force Development; and J-8 Force Structure, Resources, and Assessment.

On the night shift, before the ground war began, we took turns bringing in movies that we could show on our computer monitors in the NMCC. Usually there were mostly action-adventure movies with little plot, plenty of action, car crashes, and explosions. I could hear my brain cells dying. But every now and then, someone would bring in a good movie. Therefore, along with Kung Fu action tripe, we would occasionally get a "To Kill a Mockingbird."

When it was my turn, my contribution to the cultural advancement of Pentagon officers was the movie "The Americanization of Emily," borrowed from my sister-in-law Joan's library. Paddy Cheyevsky (sp?), a brilliant playwright and screenwriter, wrote the script. It was a black comedy loaded with cynicism about the Navy's part in the D-Day landing at Normandy in WWII. There were some wonderful pokes taken at the 'old school' Naval Academy boys. When I turned it in to the NMCC operations center, I realized that our Officer in Charge of the NMCC for the evening was a one-star Navy admiral. That made me a little concerned.

The movie turned out to be a great hit and you could hear the laughter all over the NMCC. When I went to the Ops center to retrieve it, I saw the admiral and immediately apologized if he was offended by the satirical shots taken at the Navy Academy.

This fine Naval officer, who by the end of the decade would become the four-star Chief of Naval Operations (CNO), looked at me and laughed. "No, Colonel," he said, "I thoroughly enjoyed it because I was not a Naval Academy graduate!"

I enjoyed that very much.

One Saturday night in mid-February while on watch, I was looking at the computer screen that showed where all allied shipping was deployed. After looking at the configuration of our Amphibious fleet in the Persian Gulf, I realized, having completed both courses (four weeks) of Air-Ground Task Force Planning school, that all the ships were aligned for an immediate amphibious landing.

After musing on that for a bit, I looked to the Air Force officer to my right and said, "Based on what I'm seeing here on the computer, I think tonight's the night." We all had been anticipating the jump off of the ground attack in Iraq, and my fellow watch standers concurred with my assessment.

I took a bathroom break, and as I was relieving my bladder, three-star Army Lieutenant General Kelly, the J-3 Operations Officer for the Joint Staff, appeared in the stall next to me for the same purpose. He was in full Army dress green uniform. The most obtuse observer at the Pentagon could easily figure out that three-star Army generals of his stature don't show up at the Pentagon on Saturday night in full uniform unless something serious is going down.

General Kelly and I had little in common as you can imagine, and I deemed it best not to initiate conversation. But after washing hands, I returned to my post in the NMCC and told my colleagues, "Tonight's the night."

Within hours, General Kelly was in the media briefing room telling the world's news services that the ground war phase of Operation Desert Storm was commencing.

I mentioned earlier about Marine Reserve tank units needing to be trained on the new (to them) M-1 computer enhanced tanks. One Marine Reserve tank company from Idaho, after having finished its refresher course on the new tanks at Fort Ord, CA, immediately deployed to the Persian Gulf. Upon arrival it was given its full complement of M-1 Abrams tanks and assigned to augment an active duty Marine tank battalion. The Active battalion commander assigned the Reserve tank company to be the spearhead of the main initial assault.

As it turned out, as soon as the ground offensive began, the Marine Reserve tank unit, with a major assist from Marine Engineers who had planted explosive charges to neutralize the mine fields, blasted through the front lines – and then proceeded to kill every Iraqi tank and armored personnel carrier in sight. When the Active Duty tank battalion followed up in trace, there were no more targets to be acquired. In the peaceful aftermath, the Reserve Marine tankers did what all warrior tank units do. They painted black rings around the tank barrels for every enemy tank and armored personnel carrier destroyed. I don't think the active duty guys were too happy about it as they had nothing to paint on the barrels of their tank guns.

The ground war in Iraq was brief, and thankfully with relatively few American casualties. As soon as it ended, our J-4 Logistics Reserve Mobilization Branch, once again, had to ramp itself up into high gear. It was one thing to deploy all these hundreds of thousands of troops and their equipment halfway around the planet, it was another thing to bring the whole kitten-kaboodle back home.

America, newly enamored of its military as opposed to the anti-Vietnam attitudes that, a generation earlier, had scorned the military and the honorable service of its members, now clamored to bring everyone home for victory celebrations.

We were faced with the reality of military requirements that dictated the order of withdrawal. And that created its own set of headaches.

Which units should re-deploy first? We had thousands of cities and towns across the country that wanted to bring their troops home so they could have a proper parade, and the pressure was on. This was a unique phenomenon. The military had been a castigated and untrusted institution since Vietnam. The outbreak of the Persian Gulf War saw a sea-change of attitude in America toward the military. General Colin Powell, the Chairman of the Joint Chiefs of Staff, had a lot to do with these changed attitudes. He was a calm and clear spokesman for what was being asked of the military. Public confidence grew with each of his briefings.

The military was now the most respected institution in America.

Prior to this Persian Gulf war, the average American blamed Vietnam on the unfortunate individual soldiers who, out of duty, had served in Vietnam and sacrificed so much.

That has always galled me. If you are anti-war, fine, but blame the politicians who start the trouble, not the poor sod who has to go over and clean up the mess.

So we were faced with a lot of challenges regarding the re-deployment of troops. We had a public that now found themselves profoundly patriotic and now respected the military at a level not seen since WWII. Case in point: the legendary 'Hellfighters of Harlem" was an all-black unit from New York City. The unit had originally distinguished itself in World War I when it was recruited as an all-black infantry unit (one of the few in that war) and deployed to France. Every other 'Negro' unit that had been drafted for that war had been designated to the area of "service support." That usually meant ferrying supplies and ammo to the front lines and returning with casualties on stretchers. All of those units performed with distinction and suffered huge casualties.

The Hellfighters, however, distinguished themselves in front-line infantry combat to such a degree that the French government had awarded the unit the 'Croix de Guerre" for combat gallantry.

Now in 1991, the unit had been re-designated as a "Heavy Lift" transport unit. Their motor transport support mission was to move M-1 tanks weighing 60 tons, as well as heavy artillery, and extremely heavy vehicles from ship to battlefield and back again. The heirs to the WWI Hellfighters unit had been among the first to deploy to Iraq with their huge transport vehicles and they arrived in Saudi Arabia awaiting the ships that soon discharged these heavy vehicles.

A major part of the tactical assault plan was to move a huge number of tanks 60 miles inland. Once the main assault was begun near the Persian Gulf, the tanks that had been moved inland would do an 'end-around' sweep to flank and cut off the Iraqi army preventing it from retreating from Kuwait.

It was the job of the Hellfighters to make sure that plan succeeded, and the unit did it brilliantly. But now, with the ground war ended, all those scores of thousands of vehicles needed to be cleaned, given maintenance, refurbished where necessary, and then redeployed aboard Navy and commercial maritime ships. So our heroes from Harlem were stuck there until the job was completed in August.

Meanwhile, back home at the Pentagon, General Powell was getting inundated with letters from Senators and Congressmen, as well as private citizens, wanting to know why the troops from their states or home districts hadn't been re-deployed yet. These letters were farmed out to various sections of the Mobilization Division of J-4, and the distribution was coordinated by a Navy Commander from a shop down the hall from ours.

The letter from Congressman Rangle, the US House Representative from Harlem, to General Powell asking him to intercede and bring the Hellfighters back in time for the big New York City parade in June was passed down, somehow, to me for a response. The only caveat from General Powell's office was to keep the letter, amicable, breezy, and friendly.

So I contacted the Army National Guard and asked for the history of this unit. And that's where I found out, as I mentioned above, how distinguished this unit's history had been. This was a unit with a storied past, and all the politicians, Black and White, in New York wanted the Hellfighters showcased in the big parade they planned for early June.

After doing my research, I crafted a dandy letter that got me high praise from my mentor, Army Colonel Glenn Cassidy. As a veteran of crafting letters and memos for General Powell's signature, he said that I had captured exactly the tone that General Powell looked for in his correspondence. He told me that the general wanted a down-to-earth, personal, and warm answer to be given in each letter.

We submitted the draft back up through the Navy Commander coordinating the correspondence effort, and he kicked it right back down to us for re-write.

Remember earlier when I said the Pentagon was everything you feared it was in peacetime and everything you hoped it would be in wartime? Well, we were now back to peacetime and that meant that any correspondence, memos, proposals, etc., had to be bland to the point of being narcotizing. The peacetime ethic was, 'say nothing, but use a lot of words to do so.'

When I spoke with the Commander he made it perfectly clear that taking a friendly tone in a letter to a Congressman was just not to be done. By the time he shredded my letter and reproduced it to his own standards, it was a meaningless piece of garbage. It was my understanding from someone in General Powell's office who leaked it back to Glenn, that the general was so disappointed with the final draft that he ended up writing it himself.

I like to think that his version and mine were similar.

Now I will summarize what happened when we wrote up the "Lessons Learned from the operations DESERT SHIELD/DESERT STORM,' as well as the draft of what was called 'Annex M (Mobilization) to the Joint Operations Plan.' It is a frightening description of what happens when an entire bureaucracy gets involved in drafting up a simple document.

The final effort regarding the redeployment of our troops from Iraq was the drafting of a variety of documents. The principle one was "Lessons Learned." Next on the list, as far as our department went, was a detailed draft of the ultimate re-write of Annex M (Mobilization) to the Operational Plan.

I was given that responsibility. I won't bore you with the details, but all of this activity earned me the lone personal award I ever received – the Joint Service Commendation Medal. It was awarded to me by my boss that summer at the Global War Game at the Naval War College.

I still had two more years to serve in my Pentagon billet, but things were pretty tame now and I returned to going down there once a month for my normal two-days, plus the occasional extra Active Duty day that my superiors could come up with.

Epilogue

I finished my Marine Corps Reserve service in the fall of 1993 and retired in 1994 having completed almost twenty-six years of service. Five and a-half of those were on Active Duty with the balance being all Reserve time. From June 6, 1968, when I first signed up, until April 18, 1969, I was an Enlisted man, and I am tremendously proud of that.

While my service had its share of disappointments, over all it was a fantastic adventure. If I could leave out a couple of things, I'd gladly do it all over again. Among my joys were having the opportunity to lead Marines, being selected to the Naval War College, and serving in the Pentagon during war.

I am most proud of my Marine Corps service. I am doubly proud that I could do most of it as a Marine Reservist and now privately claim the distinction of being "twice the citizen."

This obviously has not been a combat memoir. I never saw a shot fired in anger. This has been the story of a journey that took me from a college boy's dissipation and an unfocused future to a world of some amazing adventures and opportunities for education and accomplishment.

As my recruiting Gunnery Sergeant in 1968 told me, it was a chance to serve.

Made in the USA
Monee, IL
31 May 2021

69106110R00154